Antibiotics

Are they curing us or killing us?

The Author

John McKenna has a science degree and a medical degree and has been practising nutritional medicine for over 25 years. He has come out of retirement to set up The Food Clinic which deals with weight loss issues. He is the bestselling author of *Good Food, Hard to Stomach, Natural Alternatives to Antibiotics, Alternatives to Tranquillisers* and *The Big Fat Secret.* His books have been translated into most European languages.

Antibiotics

Are they curing us or killing us?

John McKenna BA, MB, ChB

Gill & Macmillan

Gill & Macmillan
Hume Avenue, Park West, Dublin 12
www.gillmacmillanbooks.ie

978 07171 4771 7

Index compiled by Eileen O'Neill
Illustrations by Keith Barrett, Design Image
Typography design by Make Communication
Print origination by Carole Lynch
Printed and bound by CPI Group (UK) Ltd,
Croydon, CR0 4YY

This book is typeset in Linotype Minion and Neue Helvetica.

The paper used in this book comes from the wood pulp of
managed forests. For every tree felled, at least one tree is
planted, thereby renewing natural resources.

A CIP catalogue record for this book is available
from the British Library.

5 4 3 2 1

I wish to dedicate this book to my granddaughter Abigail, with whom I hope to be able to spend much more time in the months and years to come.

Contents

Acknowledgements

I would firstly like to thank my editor Fergal Tobin, who retired at the end of 2013, and my commissioning editor Deirdre Nolan, who has helped a lot with this manuscript. Thank you both for all the encouragement you have provided over the past months. A very big thank you to my editor Elizabeth Brennan, who did an amazing job in getting the text ready for print. I hope we can work together on future manuscripts. In overseeing the preparation of the manuscript from submission through to printing, I would like to thank Deirdre Rennison Kunz for all her hard work.

I would like to acknowledge the help of Margo Mulligan and Trevor Peare in preparing the bibliography – thank you both very much for digging out difficult-to-find references. For providing information and answering all my queries promptly, I would like to thank Steve Payne of Wren Laboratories Ltd and Professor Laura Piddock for help provided with Chapter 9.

I would like to thank my good friend Dr Brendan McNamara for his kindness and very wise words. May life bring you great gifts. Thank you to my daughter Jackie for her assistance with publishing my first e-book and for her love and support during my illness. May marriage bring you many years of love and happiness, a few surprises and much magic. I am also indebted to my daughter Marianne for her wonderful help with my patients and for her help and support during my illness. I hope marriage also brings you many happy times, lots of fun, love and emotional security.

A big thank you to my daughter Charity for her kindness and help, especially over the past few months but also for her support over the years. You're loved and appreciated. I would also like to acknowledge my son David for his help with local talks and for his wise words and amazing insights for one so young. You have a great future ahead of you. Thanks also to my sister Pauline and her husband Enda for their help and advice throughout my life. May life bring you both great gifts.

Finally, a massive thank you to all my patients, who have been so loyal and understanding. I wish you all good health.

If I have forgotten anyone please forgive me.

Introduction

Case History: James

James was nine years old and was in the children's oncology ward in hospital. He had been diagnosed with leukaemia and was being treated with drugs (chemotherapy). He seemed to be doing well and his parents were very optimistic. Now that treatment was coming to an end there was hope that James could return home in the coming days.

That night James had a high temperature and was feverish. The doctor was called. He examined James thoroughly and could not find a reason for the fever. He ran some tests, but these did not indicate the reason for his high temperature. Because leukaemia renders many of the patient's white blood cells useless, it left James more vulnerable to infection.

The doctor asked for help from his consultant who also could not find the cause of the high temperature. They decided to ask for help from the Professor of Microbiology, who had great experience in hospital-acquired infections.

The Professor suggested it was most likely an infection, given the fact that James had leukaemia, had just finished a course of chemotherapy and was in a hospital setting where there were known to be highly resistant bugs. The Professor suggested that James be isolated and be given intravenous antibiotics immediately.

The next day, James seemed to be improving as his temperature had reduced and he seemed in better spirits. However, by the following morning, his parents expressed concern about him. The doctors decided to do blood cultures as his temperature was rising again, despite his being on antibiotics. The Professor was now gravely concerned and suggested that James was showing signs of a multidrug-resistant bacterial infection. If he was right then there was probably little that could be done.

The blood cultures showed that James had septicaemia (blood poisoning) and the bacteria showed they were resistant to all antibiotics bar one, which would usually be held back for situations such as this. The doctors immediately switched James's antibiotic to the one that the bacteria were sensitive to, even though it had a lot of side effects.

The following day, James showed signs of an improvement and everyone was relieved that he was going to pull through. By the next day he had improved further. All in all, things were looking up. His parents looked happier, though they were exhausted by the whole process. They left his side for a few hours to catch up on sleep.

When they returned later that day, the doctors were having a discussion around James's bed. His temperature had risen again.

His mother began to cry. The Professor was called again and he explained to James's parents that they must now expect the worst as it appeared that the bacteria were now resistant to all antibiotics. However, they would await the results of another blood culture, which were due the following day.

Gradually that evening and night, James deteriorated, and by the following day he was gravely ill. The blood culture confirmed the Professor's suspicions – the bacteria were untreatable. It was very likely that James would die as he had very few functioning white blood cells with which to defend himself.

Twenty-four hours later, James had passed away. The bacteria were so highly resistant that there was nothing the doctors could do. His parents were distraught with grief. They had lost a child because of a hospital-acquired infection.

Hospital-acquired infections are becoming progressively more common across the world. Cases such as James's are all too frequent now in many Western hospitals. In the US, over one hundred thousand people die per year from hospital-acquired infections such as urinary tract infections, surgical wound infections, pneumonia and septicaemia. In the UK the number of such infections is also increasing, as it is in Ireland.

As the chance of contracting an infection in hospital increases, hospitals are becoming more dangerous places. In the near future it may become risky to have simple hospital procedures carried out, such as an angiogram, because of the risk of infection with a highly resistant bug.

Bacteria develop resistance when exposed to antibiotics. They do this by altering their DNA, or genetic material, to allow them to make chemicals that protect them against antibiotics. This is Nature's way of facilitating or allowing adaption. Infections caused by antibiotic-resistant bacteria have become more widespread over recent years. Initially they were restricted to hospitals, but now they have found their way into the community as well. This is discussed in more detail in Chapter 3.

What is of most concern is that simple infections such as a sore throat or a urinary tract infection may become untreatable. Many authorities around the world are now predicting such a scenario. In March 2013, Sally Davies, the Chief Medical Officer for England, was reported as saying that antibiotic-resistant bacteria with the potential to cause untreatable infections pose a catastrophic risk to the population (Walsh, 2013).

If tough measures are not taken to control the use of antibiotics and no new ones are discovered, we will find ourselves in a health system not dissimilar to that of the early nineteenth century.

So what are antibiotics? How much do we know about them? What is bacterial resistance to antibiotics? What are the less well-known side effects of these drugs? How can we protect our bodies against infection so that we can reduce our reliance on antibiotics? These are some of the questions I shall discuss in this book. I shall attempt to do this in as simple a manner as possible, keeping technical jargon to a minimum.

This discussion of antibiotics comes at a time when many health authorities and academics around the world are voicing their concerns about antibiotics in the press and on the internet. We have been led to believe that death from an infection is a thing of the past. The above case history indicates that this is far from the truth.

Chapter 1
The Antibiotic Story

'We are losing the battle against infectious diseases. Bacteria are fighting back and are becoming more resistant to modern medicines. In short, the drugs don't work.'

PROFESSOR SALLY DAVIES,
CHIEF MEDICAL OFFICER FOR ENGLAND

Why are the authorities painting such a gloomy picture of the future? After all, antibiotics are relatively new drugs. How come they are no longer being viewed as life savers? What has gone wrong? Let us go back several decades in an attempt to answer these questions.

THE PRE-ANTIBIOTIC ERA

Most people and many doctors have no recollection of pre-1940s medicine. There was little in the way of effective curative treatments available. The medicines of the day were mostly lotions and potions, which were designed more to lift the spirits of the patient than to effect a cure.

The book *The Youngest Science: Notes of a Medicine-Watcher* by Dr Lewis Thomas describes this era very accurately. Dr

Thomas grew up watching his father, a small-town physician, administer medicine to his patients. His father instilled in him the idea that there was little that could be done about many of the ailments he encountered. Most of the potions he prescribed were placebos and contained a high level of alcohol.

Previously, opium had been the prime ingredient in these potions before its highly addictive properties were discovered and it was removed. Some potions contained quinine, strychnine and iron. Treatments included bleeding, cupping, purging and other drastic measures.

In effect, a cure was not a reality unless the body was able to heal itself. With infections there was little expectation of cure. But the discovery of penicillin was about to revolutionise all of that.

DISCOVERY OF PENICILLIN

In 1928, while attempting to grow the bacterium *Staphylococcus spp.* on an agar plate, Dr Alexander Fleming noticed that the growth of this bacterium was inhibited by a mould that had accidentally contaminated the plate. He decided to identify the mould, which turned out to be *Penicillium notatum.* Fleming was very excited by this discovery.

He cultured the mould in a special broth and injected the broth into some of his patients who had infectious diseases. The results were very encouraging and his treatment proved to be non-toxic. However, when he presented his findings to a clinical meeting in London in 1929, his colleagues in the medical profession were not particularly impressed.

It took two gifted researchers, Dr Howard Florey and Dr Ernst Boris Chain, who worked at Oxford University in the late 1930s and early 1940s, to realise the importance of Fleming's work. It was through their pioneering work that penicillin was brought into clinical use.

Florey was eager to form a group of researchers who were interested in finding effective antibacterial substances. Florey was a microbiologist and clinician, while Chain was the chemist capable of isolating, purifying and studying the properties of potential antibacterial substances. Their research team was made up of 20 of the best scientists in Britain at the time. They focused their attention on the work of Alexander Fleming and worked at purifying penicillin, studying its properties and testing its effectiveness.

In 1941, the Oxford group conducted the first clinical trial of penicillin. Their patient was a 43-year-old man who was suffering from septicaemia caused by the bacterium *Staphylococcus aureus.* His name was Albert Alexander and he was a police officer in England. He was admitted to an Oxford infirmary on 12 October 1940 with an infection on his face. The infection had begun as a small sore at the corner of his mouth and had spread over a few weeks to cover his face, scalp, eyes and neck. Multiple drainage procedures had been performed to try to drain the pus and kill the bacteria but these had not worked. His left eye was so badly infected that it had to be removed surgically. He was running a high fever.

The man was dying so Florey decided to inject a low dose of penicillin directly into a muscle every three hours for five days. Several more doses were given, and slowly but surely Alexander began to improve. The infection began to clear, his fever broke and he regained an appetite. By the fifth day of treatment he was noticeably better. Unfortunately, the supply of this experimental drug, penicillin, ran out. Alexander did well for the next ten days but then the infection returned. Shortly afterwards, he died.

Despite his death it was clear to all that penicillin was extremely effective at fighting serious infection.

The Oxford group's next challenge was to find a way to produce penicillin in large amounts. All efforts to get industrial support for their research in Britain proved fruitless and in 1941 they went to the US. Here they succeeded in getting a number of drug companies involved in the industrial production of penicillin. These drug companies made penicillin a therapeutic reality.

Subsequent clinical trials produced amazing results. Penicillin proved to be remarkably effective against a whole range of infections, including pneumonia, scarlet fever, septicaemia, streptococcal sore throat, diphtheria, gonorrhea and rheumatic fever. A general belief emerged that it was effective against any disease, especially any infection – a myth that is still prevalent today. There was tremendous publicity surrounding this new 'miracle drug', and in 1945 Fleming, Florey and Chain received the Nobel Prize in Physiology or Medicine.

Penicillin was later produced in oral form and was also added to products such as lozenges, salves, cosmetic creams and nasal ointments. Prior to 1955, its sale was not controlled and anyone could buy it over the counter without the need for a prescription. It was not known that excessive and uncontrolled use of penicillin led to the overgrowth of resistant microbes in the bowel. Microbes or micro-organisms are bugs that include bacteria, viruses and fungi. By 1955, most nations had begun to restrict the sale of penicillin, but by then the damage had been done. Resistance had become a major problem and epidemics of staphylococcal resistant bacteria began to emerge in hospitals.

OTHER FIRST-GENERATION ANTIBIOTICS

Streptomycin is another antibiotic that was developed in the 1940s. It was isolated in 1943 and was the first antibiotic to offer hope to those suffering from tuberculosis (TB). It is still used in the treatment of TB today, but its main drawback is the nasty side effects associated with it, side effects not seen with penicillin. These include deafness and kidney damage.

However, the main problem encountered in the use of streptomycin, which restricted its effectiveness, was resistance. The speed at which bacteria were able to develop resistance was a surprise to everyone. Because of this, major efforts were made to find other antibiotics.

In 1947 a newly discovered antibiotic called chloramphenicol was used in a clinical trial to treat an epidemic of typhus in Bolivia. Its success in curbing the epidemic led to its use in the treatment of typhoid fever, meningitis and brucellosis. At last scientists were discovering substances that could treat serious infections.

The euphoria that surrounded the discovery of chloramphenicol was dampened somewhat when it was shown to have serious side effects. By 1950, many investigators had become alarmed by the mounting evidence linking it with serious blood disorders including anaemia and leukaemia.

Today the use of chloramphenicol is rare in the Western world, where safer but more expensive drugs are available. It is limited to use in ear drops and eye drops. In developing countries, however, it is still widely used because it is so cheap to produce.

In 1948 in the University of Cagliari, Sardinia, researchers isolated a new group of antibiotics called cephalosporins. These new antibacterials were shown to be effective in the treatment of a wide range of infections. They destroy bacteria in a manner similar to penicillin and are valuable alternatives, especially where resistance to penicillin is a problem. The added advantage is that they have very low toxicity, although allergic reactions occur in about 5 per cent of patients.

Research into the development of new cephalosporins continues today.

By the late 1940s, yet another group of powerful antibiotics were discovered, in the US. It was called the tetracycline group of antibiotics. Today, tetracyclines rank second only to penicillin in their use worldwide.

Because they are active against a broad range of bacteria and are relatively cheap to produce, tetracyclines quickly gained favour and are now used to treat a long list of infections. They are especially popular in developing countries because they are so inexpensive.

The extensive research done on tetracyclines has shown them to be very effective, but, like many other antibacterials, they have a number of toxic side effects. Tetracyclines form complexes with calcium in growing bones and teeth, which can weaken them and also cause discolouration and damaged enamel in teeth. Tetracyclines also cross the placenta and can have a toxic effect on a growing baby. As a result, they should never be used during pregnancy. They should also not be used by young children below the age of seven.

Tetracyclines are called broad-spectrum antibiotics because they kill a broad range of microbes. Because of this they damage the gut flora, causing the overgrowth of *Candida albicans, Staphylococcus spp.* and *Clostridium difficile.* Liver and kidney damage can also occur with the use of these drugs, as can allergic reactions such as hives, skin rash and asthma.

Because the tetracycline antibiotics form complexes with calcium, magnesium and iron, they should not be taken with mineral supplements containing these minerals. It is also best not to take them with dairy products for the same reason.

Early antibacterials such as penicillin, chloramphenicol, tetracycline and the cephalosporins are today called the first generation of antibiotics. Suddenly, by the 1950s, it was possible to treat a whole host of infections. Doctors were so excited by this development that they began to speak about a time in the future when death from infection would be a thing of the past.

Unfortunately, this has not come to pass. Nature has had something to say in the interim. Bacteria have outsmarted everyone. Bacteria, which are single-celled organisms, have basically rendered most antibacterials useless. But more of this later – let's continue with the history of antibiotics.

SECOND-GENERATION ANTIBIOTICS

After the development of a number of antibiotics during the 1940s, there was no further research done until the 1960s, when a new wave of development led to a second generation of antibiotics. Among these new drugs was methicillin, a synthetic version of penicillin, which was produced specifically to overcome the problem of resistance to penicillin. Methicillin was hailed as a major breakthrough in the fight against resistance, and scientists believed they could now win this battle. Unfortunately, this optimism was short lived as bacteria quickly became resistant to methicillin too.

At about the same time, ampicillin, another synthetic version of penicillin, was developed to broaden the range of infections that penicillin antibiotics could treat. It has now replaced penicillin to a large extent. Ampicillin is often the antibiotic of first choice in the treatment of a whole range of infections, including respiratory tract and urinary tract infections. It is sold under the tradenames Penbritin and Ampicillin.

Amoxycillin is another derivative of penicillin. Like ampicillin, it has a broad range of activity as it can treat a whole host of infections. It is also very widely used today. It is sold under the tradenames Amoxil and Amoxicillin.

Later in the 1960s, gentamycin was discovered (tradename Garamycin). It is in the same group of antibiotics as streptomycin, a first-generation antibiotic. Like streptomycin, it has serious side effects, particularly as regards the ears and kidneys. As a result of this, it is generally reserved for serious infections.

THIRD-GENERATION ANTIBIOTICS

More recently, in the 1970s and 1980s, we have seen the development of a whole range of new antibiotics called fluoroquinolones. These are all synthetic drugs derived from the anti-malarial drug chloroquine. Hundreds were developed but only a few are in clinical use. The names of these drugs usually end in 'oxacin', such as ciprofloxacin (tradename Ciproxin or Ciprobay).

You may be aware of these drugs from news stories concerning lawsuits brought against the drug companies that manufacture them. For example, the numerous class action lawsuits against Johnson & Johnson, who manufacture the drug levofloxacin (Levaquin). These lawsuits have to do with the dangerous side effects of these drugs.

Although these drugs are very effective and useful in limited situations, their nasty side effects have severely curtailed their use in medicine. They are quite widely used in veterinary medicine, but governments are trying to restrict even this usage due to the speed at which bacteria can develop resistance to them.

The search for new antibacterial agents continues. The pace, however, has slowed remarkably as it is now much more difficult for drug companies to get approval for new drugs. The time delay between the discovery of an antibiotic in the laboratory and the approval to produce it commercially is so great that it has led many companies to abandon the search for new antibiotics completely.

It is also true to say that the drug companies are also discouraged by the pace at which bacteria can develop resistance, which can render a new drug useless in a short space of time. Add to this the fact that many antibiotics, such as the fluoroquinolones mentioned above, have nasty side effects and you can begin to see the barriers facing drug companies.

Antibiotics are relatively new. They have only been in use for over fifty years. We have had three generations of antibiotics to date and they have done much to improve survival rates for those with serious infections. They have indeed been life-saving drugs.

However, even during the early stages of antibiotic development, it was clear that some bacteria were able to survive and multiply in the presence of antibiotics. These bacteria had acquired resistance. In his Nobel lecture in 1945, Dr Alexander Fleming said:

> The time may come when penicillin can be bought by anyone in the shops. Then there is the danger that the ignorant man may easily under-dose himself and by exposing his microbes to non-lethal quantities of the drug make them resistant.

In his Nobel lecture Dr Fleming also predicted that this problem of bacterial resistance would worsen if penicillin was made available in oral form, if inadequate doses were given, if a course of treatment was not completed or if people were given a course of treatment that was too long.

Many of us, I am sure, have skipped a dose or two in a five- or seven-day course of these drugs. We have ended up with a few extra capsules, which we stored away in our medicine cupboard for weeks or months. Perhaps on occasion we stopped our treatment once the symptoms abated, again ending up with an incomplete course of treatment.

These very common scenarios are exactly what Fleming was warning us against. He was predicting the downfall of these drugs and how right he has proven to be. So, from the outset, people saw that bacteria would be able to counteract antibiotics. Few, however, foresaw just how big a problem this would become.

Bacterial resistance is now approaching a nightmarish scenario. From the 1950s until recently we have been dealing largely with hospital-acquired infections, which are resistant to many antibiotics and disinfectants. For example, most people are now familiar with the hospital infection MRSA (methicillin-resistant *Staphylococcus aureus*). I am sure you have heard stories of hospital wards and surgical theatres being shut down because of this bug.

In the early 1980s, a number of hospitals in Melbourne, Australia were plagued by infections that were resistant to almost all known antibiotics. The bacterium responsible for these infections was MRSA. The bacterium was resistant not only to antibiotics but to many disinfectants as well, making it virtually impossible to kill. Only one antibiotic remained effective, vancomycin, which is both expensive and toxic. Doctors had no choice but to use it and by doing so they eventually got the outbreaks of MRSA under control.

Then, in May 1996, a four-year-old child was admitted to hospital in Japan for open heart surgery. After the operation the child developed an infection at the site of the surgical wound. The infection was caused by MRSA. When tested in the laboratory, the bacterium proved resistant to vancomycin. Alarm bells rang across the world and government health departments began to panic. The bug had become resistant to vancomycin, the last line of defence. We now had a 'superbug' on our hands.

Doctors cleaned the child's wound daily and used a variety of drugs to try to kill the bug. To everyone's amazement, the child survived the infection. This was a narrow escape from disaster. Everyone involved breathed a sigh of relief.

Later still, in 1997, strains of MRSA that were resistant to vancomycin began to appear in hospitals in the US. In May 1998, an elderly patient in a New York hospital died of an infection caused by this superbug.

Needless to say, many more alarm bells began to ring and the World Health Organization warned doctors to stop over-prescribing antibiotics.

Sadly the warning has gone unheeded and today we are now faced with the prospect of minor infections, such as a sore throat, becoming more difficult to treat. In the not-too-distant future, as antibiotics become more ineffective, these community-based infections may become untreatable. However, with some prudence on the part of patients and doctors, it is possible to prevent this from happening.

Chapter 2

What We Know about Antibiotics

Antibiotics are among the most commonly prescribed drugs in most Western countries. We have all used antibiotics at some point in our lives – for sore throats, ear infections, dental abscesses and so on. Usually a doctor will prescribe them for a short period of time – a five- or seven-day course. In some cases, antibiotics may need to be taken for a long period of time, months or years. Among my patients, many had taken antibiotics for years to treat acne.

There has been a lot of publicity in recent years about the use and misuse of these drugs and the role they are playing in the outbreak of serious hospital infections such as *Clostridium difficile* and MRSA. But how much do we know about these drugs? Are they being used correctly and why are they being blamed for the closure of hospital wards and surgical theatres?

Unfortunately there is quite a lot of ignorance about infectious diseases and the role of antibiotics in keeping them at bay. Among the general public there is a belief that, because of the availability and effectiveness of antibiotics, people do not die from infections, at least in the Western world. There is a misconception that only people in the developing world still die of infection because they do not have the necessary drugs or medical equipment. However, World Health Organization statistics reveal that infectious diseases are the second leading cause of death worldwide, killing over fifteen million people per year. In the Western world, many of these infections are contracted in hospitals and care centres, and are especially prevalent among the elderly. In 1969 the Surgeon General of the US Dr William H. Stewart declared: 'It is time to close the book on infectious diseases and declare the war against pestilence won' (Spellberg, 2008). With statements such as these coming from those at the top of the medical profession, is easy to see why misinformation has built up around the effectiveness of antibiotics through the decades.

There is also the perception among the public that hospital-acquired infections only occur in 'unclean' hospitals, where safe hygiene procedures are neglected. While hygiene is very important, it is not the underlying cause. As mentioned, the reason why hospital infections have become so prominent is because of the misuse of antibiotics and the subsequent development of resistant strains of bacteria. Second, there has been a huge increase in the number of invasive procedures undertaken in hospitals in recent years, creating more opportunities for infection to take hold.

Over the years, various pieces of research have been carried out to determine what the public knows about infections and the best way to treat them. Let us now look at the results of some of these.

To test what the general public knew about antibiotics, a survey was carried out by the Mater Children's Hospital in Brisbane, Australia in 1975. Survey participants were asked to look at six statements and to indicate if they thought the statement was true (Chandler and Dugdale, 1976).

Here are the results of the survey:

1. *Antibiotics kill viruses* – 55% thought this was true.
2. *Antibiotics kill bacteria* – 46% thought this was true.
3. *Antibiotics are a stronger form of aspirin* – 13% thought this was true.
4. *Penicillin is not an antibiotic* – 15% thought this was true.
5. *Antibiotics are good for colds and flu* – 75% thought this was true.
6. *Antibiotics are good for gastroenteritis* – 40% thought this was true.

Let's examine each of these statements and responses individually:

1. *Antibiotics kill viruses*: a staggering 55 per cent of people believed that antibiotics kill viruses. Antibiotics do not kill viruses; they kill bacteria. If we take those surveyed to represent the general public, more than half the population believe erroneously that antibiotics kill viruses and treat viral infections.
2. *Antibiotics kill bacteria*: less than half of the participants in this survey agreed with this statement, which is true.
3. *Antibiotics are a stronger form of aspirin*: generally people got the answer to this one correct; only 13 per cent were misinformed. We can take from this that most people realise there is a difference between these two types of drug.

4. *Penicillin is not an antibiotic*: again most people got this right. Penicillin is of course an antibiotic. Only a small percentage of the participants thought it was not an antibiotic.
5. *Antibiotics are good for colds and flu*: three-quarters of those surveyed got this one wrong. There is clearly a misconception amongst the general public that antibiotics are effective in the treatment of colds and flu. Colds and flu are caused by viruses and so antibiotics are completely ineffective against them. However, doctors often prescribe an antibiotic if you have flu 'in case' the infection progresses and turns out to be a bacterial infection. This is bad medicine and has had serious consequences for all of us.
6. *Antibiotics are good for gastroenteritis*: it is often quite dangerous to use an antibiotic to treat gastroenteritis, unless you have already established through tests that the disorder is bacterial. Then and only then should an antibiotic be considered. In the survey, 60% of people got the response correct, which is encouraging.

As you can see from the above survey, in general, the public is not well informed about when an antibiotic is necessary and when it is not. There is a clear need for education in this area.

Unfortunately it is not only the general public that has problems knowing when to use an antibiotic. The *Journal of the American Medical Association* carried out an assessment of the diagnostic accuracy of doctors in the US (Poses *et al.*, 1985). Over a period of 6 months, 308 patients complaining of a sore throat were assessed by medical doctors. The doctors were then asked if they thought the patient had a streptococcal sore throat and, if they did, to indicate what treatment should be given. The doctors suggested that 81 per cent of the patients had a streptococcal throat and many prescribed an antibiotic. The throat of each patient was swabbed and the swabs sent to the laboratory to see if *Streptococcus* was present. The laboratory results found this bacterium in only 5 per cent of the patients. The suggestion here is that doctors find it hard to distinguish between a viral and a bacterial sore throat and so they tend to err on the side of caution and prescribe an antibiotic. The results of this research indicate that antibiotics are being prescribed inappropriately; they suggest that actually the majority of prescriptions for antibiotics are unnecessary.

A similar assessment was carried out in Canada and reported in the journal *Canadian Family Physician* (Wong and Tiessen, 1989). The doctors participating in the study diagnosed 50.5 per cent of the patients with streptococcal infection and prescribed an antibiotic, whereas the laboratory results showed that only 13.5 per cent of the patients actually had a streptococcal infection. Again, the main conclusion to draw from the study is that many unnecessary prescriptions for antibiotics are being written in Canada.

Medical knowledge about antibiotics in the US was perhaps more spectacularly revealed in the form of the televised programme *National Antibiotic Therapy Test* in 1975. No less than 4,500 medical specialists took part in this test. Written case histories were presented to the doctors and they were asked what treatment they would suggest in each case. Here are the results of the test. Remember that the pass rate for most medical exams in the US is 75 per cent.

Surgeons scored	10%
Family doctors scored	15%
Obstetricians scored	18%
Pediatricians scored	24%
Internists scored	38%
Infectious disease experts scored	83%

Only those doctors who specialised in infectious diseases actually passed the test. This is very revealing and an important point to keep in mind when I discuss ways of controlling superbugs such as MRSA in Chapter 11.

A study in the *New England Journal of Medicine* further proves that antibiotics are being over-prescribed (Jaffe *et al.*, 1987). The study asked the question: Is early antibiotic treatment useful? It looked at 955 children under the age of three who were admitted to hospital with fever. The conclusions of the study were that antibiotics were given needlessly in over 90 per cent of cases of children with fever and, consequently, antibiotics are not advisable for most children with a high fever.

As you can gather from the above, doctors do not have all the answers. Indeed, doctors can often be wrong. This may make for uncomfortable reading for doctors, but it is also an uncomfortable truth for patients and parents of young children. I'm sure it would unnerve most people to realise that their doctor may very well misdiagnose an infection a lot of the time. It is important to remember that not only does the over-prescription of antibiotics lead to the growth of resistance to bacteria, but antibiotics can have numerous nasty side effects that can harm the patient (see Chapter 10). Since bacterial infection is genuinely hard to distinguish from viral infection, tests, such as taking a swab and sending it to the laboratory, are important. It takes the guesswork out of the picture and assures both doctor and patient of the need to use an antibiotic. In certain cases, it may be necessary to enlist the help of a specialist in infectious diseases.

The major international medical organisations, including the World Health Organization and the World Medical Association, have repeatedly called for more caution in the use of these drugs by all doctors. The scale of the bacterial resistance problem is of major worldwide concern. Microscopic organisms have an incredible ability to counteract and defend against any antimicrobial substance, be it disinfectants, anti-malarial drugs, antibiotics or antivirals. These micro-organisms can develop defences against anything we throw at them and do so with alarming speed. They can also pass resistance from one to the other.

We know that when antibiotics were developed after the Second World War they were hailed as miracle drugs, and everyone believed we would see the end of the scourge of infections. We believed that a new infection-free era would begin and life would improve; nobody would die of an infectious disease in the future. Unfortunately, this rosy future has not materialised. We have not only failed to rid the planet of infectious disease, but we may be losing the war against microbes to the point where hospitals are becoming unsafe places and some infections are becoming untreatable.

So what is the place of antibiotics in medical treatment? Well, we know that they are effective at treating infections caused by bacteria such as *Streptococcus pyogenes*, which causes sore throats. They act by either killing the bacteria or controlling their growth sufficiently to allow your immune system to fight the infection. As mentioned, they are useless at fighting infections caused by viruses, such as flu or the common cold.

There are instances where antibiotics can be life-saving. My own son's story is a perfect example of this.

When he was quite young he got chickenpox. He scratched a lesion on his abdomen and developed peritonitis. He had to be admitted to hospital and have an emergency operation to control the infection. He was then placed on antibiotics for the next two weeks and fortunately he survived. Antibiotics play an important role in medical treatment. They play such an important role that they should be reserved for only the most severe cases.

The consequence of the indiscriminate use of antibiotics is high levels of resistance amongst the bugs we are trying to kill. Every time we develop a new antibacterial drug, the bugs counteract it, making it ineffective or useless. This has now led us to an impasse. If doctors do not change their attitudes to these drugs, we are going to regress to the pre-antibiotic era.

Chapter 3
Bacterial Resistance to Antibiotics

Bacterial resistance to antibiotics is not a new phenomenon. Actually, it has been around for as long as bacteria themselves, but at a very low level. Bacteria and fungi coexist in soil. This coexistence is not entirely peaceful – in fact, fungi and bacteria are constantly battling for space and resources in the soil. Fungi compete with bacteria by producing antibiotics. You may remember from Chapter 1 that most of the first generation of antibiotics were isolated from fungi. In order to survive, bacteria devised a means of protecting themselves against these antibiotics – they developed resistance. Resistance can therefore be viewed as a survival mechanism.

But if resistance has always been a characteristic of bacteria, why has it now become so widespread and problematic? The answer to this question lies in the way we have approached the use of commercial antibiotics.

Case History: Mark

Mark was 12 years old and I had been treating him for recurrent ear and throat infections for a few months. I was abroad when he developed a high fever, a sore throat, and aches and pains. His mother brought him to the local doctor who diagnosed a possible viral infection and prescribed a course of antibiotics 'just to be on the safe side'. Fortunately, Mark's mother did not fill the prescription. Instead, she sought advice from my nurse, who recommended antiviral measures including taking echinacea and frequent high doses of vitamin C. This treatment worked very well, and within 48 hours Mark was back to normal.

In Mark's case, antibiotics were not needed as the child responded well to antiviral treatment. However, it is an example of a difficult situation for both doctor and parent. The doctor has to use guesswork to decide if the child's infection is viral or bacterial; the parent is often frightened that her/his child will become seriously ill. Guesswork and fear can result in inappropriate action. With an infection, it is much wiser to wait and see if simple, natural measures work. It is also much wiser to take a swab of the infected area and send it to the laboratory, in order to establish if the infection is bacterial and subsequently which antibiotic will work.

The following case history shows how self-medication with the cooperation of a pharmacist can also result in the inappropriate use of antibiotics.

Case History: John

John was the financial director of a large computer software company. He had recurring symptoms of tiredness and a sore throat for over a year. A friend of his was a pharmacist, so to save time and money John got antibiotics directly from his friend and treated himself. By the time he sought medical help, he had taken 7 courses of antibiotics in the space of 12 months. His symptoms were not only unimproved but were now constant, with the result that he found it difficult to keep working.

This man was 'too busy' to seek medical help early on. Self-treatment seemed like a good option at the time. If you self-medicate with potentially dangerous drugs, you only encourage the development of bacterial resistance in your body. This can cause long-term illness. Don't put your pharmacist in a difficult position, even if they are a relative or friend. If you are not feeling well, consult your doctor.

John's symptoms were due to a disturbance in the bacterial flora, the 'good' bacteria that protects the body. By simply altering his diet, and taking a good probiotic supplement and high-dose antioxidants such as vitamin C, he recovered his health and was able to resume work. I cannot overstate the importance of diet, especially for those living in modern cities. It is easy, with a busy lifestyle, to end up eating unnatural highly processed foods. But when the inevitable occurs and the body breaks down, we tend to reach for unnatural medicines to treat it.

In the hundred years from 1850 to 1950, there was a 90 per cent drop in the death rate in the UK. This rapid decline had nothing whatsoever to do with antibiotics, since they had not yet been mass-produced before the 1950s. The main influences on this decline were:

- Better nutrition
- Safer water
- Good sanitation

These are the very things that create a healthy body and a healthy society. A healthy body is better able to fight infection and to prevent infections from developing in the first instance. Most people believe that antibiotics treat infections. The truth is rather simpler. If we all ate natural foods, drank safe water and made our lives simpler to reduce stress, we would be a lot stronger and avoid infections more easily.

John's case history is interesting for another reason. It shows that it is not only doctors and patients who are at fault when it comes to over-prescribing – pharmacists are too. An interesting piece of research from Spain indicates precisely this. Researchers asked two actors to visit a number of pharmacies. The actors were told to pretend that they were suffering from a sore throat, bronchitis or a urinary tract infection. Antibiotics were sold in almost half of the cases. This research was reported in the journal *Clinical Infectious Diseases* in 2009 (Llor and Cots, 2009). In certain European countries antibiotics are for sale over the counter in pharmacies. You can buy them like you can buy paracetamol in Ireland.

THE ROLE OF THE DRUG COMPANIES

I spent most of my working life in different parts of Africa and I saw that it was possible to buy any pharmaceutical drug without a prescription in the marketplace in most countries that I lived in. This included South Africa, which has a huge black market in illicit and licit drugs. Worse still, I witnessed the use of drugs that had been banned or discontinued in Europe and the dumping of old stocks of antibiotics for use in African hospitals. For example, I was shocked to witness with my own eyes the drug thalidomide being used in Africa. Thalidomide causes congenital deformities if used in pregnancy. The drug company claimed that there were written warnings on the drug's label that it should not be used during pregnancy. However, the company was conveniently ignoring the fact that the drug was being sold to people who could not read.

In 2007, the *Guardian* newspaper reported the details of an illegal trial of an antibiotic in the city of Kano, Nigeria, which caused a number of deaths. The US pharmaceutical company Pfizer conducted a trial of the drug Trovan, a new antibiotic that they had developed. It is one of the family of antibiotics called quinolones, which are known to have nasty side effects. Nigerian officials say that, of the 200 children treated with the drug during an outbreak of meningitis, 50 died and many others ended up with permanent brain damage. Pfizer says only 11 children died. There was a public outcry in Nigeria and, as a consequence, the government launched an investigation. This investigation concluded that it was 'an illegal trial of an unregistered drug' and that it was 'a clear case of exploitation of the ignorant'. The Nigerian government then sued Pfizer for $7 billion. The Kano State government also sued Pfizer (McGreal, 2007).

This is quite typical of the way in which drug companies treat third world countries. They treat them with disdain. The main objective of these drug companies is to make money so that their share price on the stock market will rise and their investors will continue to invest.

In my opinion, the same drug companies are partly responsible for the alarming rise in bacterial resistance. They exert pressure on drug reps to encourage doctors to prescribe their drugs (Blumenthal, 2004). In many countries they bribe doctors with holidays, parties, presents, money for research, money for lecture tours, company shares and so on (Campbell, 2007). They do not care about the alarming rise in bacterial resistance; they are blindly focused on profits.

When you see how quickly doctors prescribe a drug instead of opting for more conservative treatment, you have to question whether the relationship between drug companies and Western medicine has become too co-dependent. Greed for money and power has brought shame and discredit to both. Take, for instance, the emerging information regarding the unsavoury relationship between drug companies and medical research.

In the 1970s there was little connection between drug companies and medical research. Today, two-thirds of all academic medical centres in the US have equity interest in companies that sponsor research within these same universities or colleges. Most heads of department now receive money from drug companies to support their department. Sixty per cent of these heads of department receive personal income from drug companies (Bekelman, 2003).

A Senate committee chaired by Senator Grassley has been established in the US to examine this too-cosy relationship between the drug companies and medical research. This committee has investigated serious cases of corruption. One such case involves Joseph Biederman, Professor of Psychiatry at Harvard Medical School, who is accused of treating children as young as two with a cocktail of powerful drugs, many of which have not been approved by the US Food and Drug

Administration (FDA). The drug companies that manufacture these drugs paid him $1.6 million. The hearing is ongoing.

Professor Alan F. Schatzberg, Chair of the Department of Psychiatry at Stanford Medical School, was also investigated by the same Senate committee. He had over $6 million worth of stock in a drug company called Corcept Therapeutics and was the principal investigator in a number of clinical trials involving drugs manufactured by the same drug company.

Cases such as the above discredit the medical profession, despite the fact that many doctors abide by the Hippocratic Oath. They serve to highlight that, in the overall picture of the over-prescription of antibiotics, we would do well not to ignore the influence of the drug companies.

THE USE OF ANTIBIOTICS IN FARMING

Antibiotics have been used in animal feed since the 1950s. It was discovered that animals that were given low levels of antibiotics in their food showed significantly more weight gain compared to animals that were given normal animal feed. The whole of the animal husbandry industry latched onto this information, as did the drug companies. Today, 80 per cent of antibiotics sold are used in animal feed (Falkow and Kennedy, 2001).

In Europe, penicillin, tetracycline and streptomycin were banned for use as growth promoters in livestock in the 1970s, and by 2006 all antibiotic growth enhancers were banned. However, the situation is quite different in the US.

For many years now the FDA has been aware of the possible dangers of using antibiotics in farming, but because it is a political body it is subject to political interference. Food companies that produce meat and drug companies that supply the antibiotics have a powerful political lobby and so far have succeeded in preventing a ban on the use of these drugs. However, the FDA have had a lawsuit brought against them by a group called the Natural Resources Defense Council (NRDC), one of the most powerful environment protection groups in the US (www.nrdc.org). This lawsuit was filed in May 2011 in an attempt to finally end the use of antibiotics in animal feed.

Most of these antibiotics are given to healthy animals to make them bigger. Chickens, for example, are often injected with antibiotics from day one until they are slaughtered. This is to increase their body weight and to speed up their growth. The faster the chicken grows the sooner it can be slaughtered, thereby increasing profits.

In more recent times, because of intensive farming methods such as battery farming, even more antibiotics have been used on pigs, turkeys, chickens and cattle, since they are kept in crowded, unsanitary and stressful conditions, and hence are more prone to infection. As a result, such animals have become breeding grounds for highly resistant superbugs. Meat produced from these animals can carry bacteria with extremely high levels of resistance, which may then get passed to the consumer.

Reports from the FDA indicate that a high percentage of the meat for sale in supermarkets contains multidrug-resistant bacteria. In Canada, researchers have blamed the bacteria found in meat for the persistent bladder infection in women reported by ABC News on 11 July 2012. Dr Amee Manges of McGill University, Montreal suggested that the bacteria causing the bladder infection in these patients matched the bacteria in the meat for sale in supermarkets and grocery stores, specifically chicken meat. The DNA fingerprint matched exactly. This was the first time a direct link was made between superbugs affecting humans and chicken meat. The researchers suggested that over eight million women may have been at risk of contracting a difficult-to-treat infection as a result of eating these chickens (Avila, 2012).

That meat can be contaminated by superbugs is not a new revelation. Food has been constantly monitored for this by the FDA in the US and the European Food Safety Authority (EFSA) since the early 2000s. What is new is the proof through DNA fingerprinting that these bugs may now be causing infections in humans. I anticipate further research to confirm this from Montreal.

Although, as mentioned, there are restrictions in Europe as regards the use of antibiotics in farming, antibiotics can still be used to treat infections in livestock, and it is common knowledge that some farms circumvent the legislation in this way. Antibiotics should be administered by a veterinary surgeon directly or through writing a prescription that a pharmacy can fill. Only in this way will the indiscriminate use of antibiotics stop. These drugs have been sold over the counter to farmers for too long.

The use of antibiotics as growth promoters must be banned worldwide if there is any hope of getting the levels of bacterial resistance down.

THE HUMAN FACE OF RESISTANCE

There has been much publicity about outbreaks of MRSA in areas of the community in several countries, for example, among army recruits and amateur and professional athletes. In the US, outbreaks have occurred among players on major college football teams and among professional footballers on teams such as the St Louis Rams. These outbreaks may have to do with the fact that army recruits and athletes playing contact sports can easily get cuts, nicks and bruises, which allow the bacteria entry into the body.

What is of even more concern is when healthy people in the community, with no evidence of cuts or bruises, contract MRSA. This has become so common on a global level that authorities and experts are warning that everyone is at risk.

In October 2007 it was reported that a healthy high school student called Ashton Bonds, from Virginia in the US, had died after contracting MRSA two weeks earlier. There was a public outcry about his death and as a consequence the state government closed 21 schools for a week to allow for thorough cleansing.

A number of similar cases have occurred across the US. For example, in Pennsylvania 13 members of a high school football team were diagnosed with MRSA infections. The sad cases are the ones that prove to be resistant to all the available antibiotics. Healthy young people in the prime of their lives die because of bacterial resistance. This is going to become commonplace if action is not taken urgently.

The Infectious Diseases Society of America, an international body of professionals, has been gathering examples of these sad cases on its website (www.idsociety.org) to raise public awareness of the problems being faced by clinicians on a daily basis. There is the story of Rebecca, a 17-year-old healthy student, who developed pneumonia out of the blue, was admitted to hospital and died shortly afterwards, despite having the best medical care available. She was diagnosed as having multidrug-resistant MRSA.

Doctors are shocked by the patterns of illness they are seeing. Up until the early 1990s, most superbugs were contracted by patients admitted to hospital for an operation or a procedure. Hospitals were thought to be breeding grounds for these superbugs. Since the early 1990s, however, the pattern has changed and most superbug infections are contracted within the community.

This pattern of resistance that begins in hospitals and ends up in the community is exactly what was seen in the 1950s with penicillin resistance, as Figure 1 indicates.

Figure 1: The Pattern of Resistance that Begins in Hospitals and Ends Up in the Community, as Seen in the US

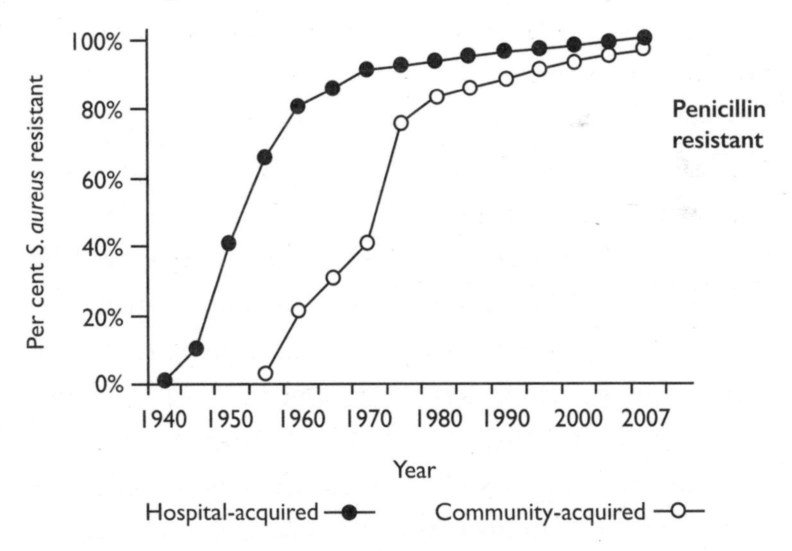

Source: Chambers, H.F. (2001), 'The changing epidemiology of *Staphylococcus aureus?' Emerging Infectious Diseases*, vol. 7, no. 2, pp. 178–82. Adapted with permission.

Most of the initial cases of drug resistance are seen in in-patients who are admitted to hospital for some other reason and who then contract a hospital-acquired infection. After some years, the superbugs move out of the hospital and infect people in the community. Then most of the cases seem to be community based.

Other patterns can also be observed. Superbugs are becoming more resistant to antibiotics with time, making them more difficult to treat. Usually the antibiotic vancomycin is held in reserve for MRSA infections, but now the bugs are displaying resistance to this drug as well.

The other pattern being noted is that it is the young – the above stories testify to this – and the elderly, especially those in care homes or other institutions, who are most at risk. Many of the case histories, particularly those that receive media attention, involve young healthy teenagers who contract the infection out of the blue.

HOW BACTERIA DEVELOP RESISTANCE

The mechanism by which bacteria overcome antibiotics is fascinating. You can only be in awe of these clever microbes and the ways in which they can outwit and nullify our efforts to kill them.

In a sense, it is our narrow-minded and arrogant belief that we can control Nature by killing off what we interpret as unnecessary or harmful that is tripping us up. In actual fact, the shoe is on the other foot. Infectious diseases have the potential to control *us* and wipe *us* out.

Our limited thinking and our lack of awareness are now forcing us to view things differently. We must accept that even pathogenic, disease-causing bacteria have a positive and important role to play in Nature. We don't have to understand what this role is; we need only respect it. And respect is the key to solving the problem of bacterial resistance.

Nomadic people such as the Masai of East Africa have always had this respect. They have an understanding of the interconnectedness of all living things and so are always attempting to work with Nature by obeying its laws. These people do not see themselves as different from, or better than, the rest of Nature. Many Westerners, on the other hand, see humans as all-important, as separate in some way, and therefore able to rise above Nature and control it. Single-celled microbes called bacteria have taught us the folly of our ways. Not only have they proven themselves able to evade our magic bullets – antibiotics – but they are teaching us a valuable lesson.

In truth, we all need to learn this lesson and to shift our thinking away from control – control of Nature, control of people, control of land and control of money – towards living in harmony with everyone and everything around us. If we were in tune with Nature we would understand that all forms of life, including bacteria, play a very important role in Nature. If we try to exterminate what we consider as bad or harmful, we are upsetting the natural order. By trying to exterminate certain bacteria, we are in fact encouraging these bacteria to survive.

Let's now examine the ways in which bacteria are able to fight back.

Mutation

Bacteria have been able to adapt to hostile environments and survive over the centuries through a process of mutation. Genetic material can mutate or change to produce a gene. This gene can help the bacteria to survive in the face of any toxic stuff in the environment, including antibiotics. At low levels of antibiotic usage, this is all that is necessary for the bacteria's survival. The presence of an antibiotic kills off the susceptible bacteria, which do not have the new gene, and favours the growth of those bacteria that have the mutant gene and so are not harmed by the antibiotic.

As you read earlier in this book, Dr Fleming noticed these mutant forms in his experiments and warned about them. He predicted that the more widespread the use of antibiotics, the more widespread and numerous these mutant forms would be.

Plasmids

While we were manufacturing antibiotics, using them freely and patting ourselves on the back for our wonderful achievements, bacteria were busy developing a more efficient means of protecting themselves. They developed new and improved methods of survival in the form of plasmids. I first learned about plasmids in the early 1970s at university, but little did I realise that they would assume such huge importance today.

Plasmids are extra bits of genetic material, or mini-chromosomes, which are found in bacteria. Plasmids are produced in response to changes in the environment of bacteria. They contain new genes to help the bacteria adapt faster.

Plasmids are constantly changing. They continually lose

genes that are no longer of use and acquire new genes. The environment dictates and selects those genes that are of value and need to be retained, and those that can be discarded.

Through our misuse of antibiotics, we have encouraged the development of plasmids and ensured their continued importance. The main function of plasmids is to prevent bacteria from being killed by antibiotics. Plasmids were unknown until the 1970s, when resistance became a major problem. They rang the death knell of penicillin and warned of even bigger problems to come.

A unique characteristic of plasmids is the fact that they can be transferred from one bacterial cell to another; they can also be transferred from one species of bacteria to another, such as from *Staphylococcus aureus* to *Escherichia coli* (E.coli). This is why all types of bacteria can develop resistance so rapidly. This is also the reason why bacteria can become resistant to many antibiotics simultaneously, a characteristic that is known as multiple drug resistance.

MULTIPLE DRUG RESISTANCE

Something very startling happened in a hospital in Japan in the late 1950s – the birth of multiple drug resistance. In this hospital, a number of patients were suffering from dysentery. The bacteria causing the infection were resistant to a number of antibiotics, including tetracycline, streptomycin, chloramphenicol and the sulphonamides (Septrin; Bactrim). Multiple drug resistance was unknown prior to this. Now it sent shock waves across the planet.

By 1966, a number of countries had reported multiple drug resistance. In one South African hospital, 50 per cent of the E.coli bacteria (a normal constituent of the gut) isolated from the faeces of patients showed resistance to more than one antibiotic. Sharing genetic information is the name of the game in the world of microbes, especially if their survival is threatened in any way. Certain resistance information was carried by the plasmids inside the bacterial cells. These plasmids were then transferred to other bacteria, making these multi-resistant as well.

Today, multidrug-resistance has become a worldwide issue. In the words of the Chief Medical Officer for England Professor Sally Davies, 'it is a bigger threat than terrorism' (Davies, 2013, p. 73) (Walsh, 2013). Today, virtually the whole planet has a problem with high levels of resistance, not just to anti-bacterials but to antiviral, antifungal and antimalarial drugs. The problem is not limited to the developed or developing world; it affects and unites us all.

THE SOLUTION

It is possible to solve the problem of resistance. Before discussing a possible solution, there is one more thing you should know about plasmids. So far, you have learned that plasmids are a smart way of allowing bacteria to survive in a hostile environment, such as in the presence of antibacterial drugs. However, if a bacterial cell contains one plasmid or more, this can be a disadvantage to the cell in two ways.

First, carrying extra genetic material such as plasmids can consume a lot of the cell's energy, so that less energy is available for growth and reproduction. Second, carrying these extra genes will make the bacterial cell less virulent; in other words, too many of the cell's resources are consumed by ensuring survival. If I spend most of my time and energy growing vegetables so that I have food to survive, I will have little time and energy to write books and see patients. My energy will be consumed by a more basic need – the need to survive.

We are putting pressure on bacteria to carry these extra genes through our gross misuse of antibiotics. If this pressure was removed by withdrawing the use of antibiotics for a period of time, bacteria would then discard their plasmids and return to their original state. Herein lies the solution.

In one hospital in Cape Town, South Africa, the doctors had a problem with bacterial resistance to one particular antibiotic called gentamycin. By using an alternative antibiotic and banning the use of gentamycin for a period of five years, the particular species of bacteria, *Klebsiella pneumoniae*, lost its resistance genes and once again became sensitive to gentamycin. This antibiotic became useful again in the treatment of infections caused by *Klebsiella pneumoniae.*

This story offers a ray of hope. It suggests that a more prudent approach to the use of these drugs will indeed result in bacterial changes. These changes will lead to reduced bacterial resistance and a return to the natural state of bacterial sensitivity to these drugs.

In other words, we have to stop using certain important antibiotics for a limited period to allow bacteria time to lose their plasmids. Then we can reuse these same antibiotics but in a much more controlled and disciplined way. This plan, however, would require not only the agreement of national governments but international agreement if it is to succeed.

THE CONSEQUENCES OF MULTIDRUG-RESISTANCE

As with all situations in life, we can choose to view things negatively or positively. The negative view of bacterial resistance is evident in certain literature and broadcasting, particularly from the US. Some doctors portray the problem as a having the potential to wipe out the whole of humanity in a short space of time. There is some truth in this, but it is not the full story. I believe that since we humans have created the problem, we can surely solve it too.

The more positive view of resistance is to see it as a blessing in disguise. It is a blessing in that it makes us stop and think about our actions. It makes us more aware of the consequences of those actions. It sheds light on our quick-fix approach to medicine and the impasse we have created as a consequence of this approach. In other words, it teaches us to be wiser and to delay or avoid the use of antibiotics, reserving them for the more serious cases only. It forces us to become more aware of the dangers of antibiotics and of all drugs, and of the flaws in the drug-only approach to treating patients. Resistance has prompted us to find more natural means of treating simple infections, such as the use of high-dose vitamin C.

In a broader context, the problem of resistance challenges our view of ourselves and our world. We can no longer view ourselves as wiser or more knowledgeable than single-celled microbes. Our concept of control, not just control of Nature but of many aspects of our lives, is being challenged. Resistance challenges us to be more in tune with ourselves and in so doing makes us more aware, more sympathetic and more empathetic.

Control of anyone or anything never works. When we glimpse the beauty at work within bacteria, humans and the whole of the natural world, we begin to realise that control prevents this beauty from showing itself.

The fact of bacterial resistance is helping us to find more natural means of treating simple infections, such as the use of high-dose vitamin C. My book *Natural Alternatives to Antibiotics* (2003) aims to encourage people to use simple natural modes of treatment and to only use an antibiotic as a last resort. However, I am not against the use of antibiotics as they have saved many lives; I am against their misuse and the idea that they are the only way to treat an infection. I stand for a more enlightened approach, not just to the treatment of infections but to treatment of all ailments. I stand for a more inclusive form of medical education – to include herbalism, homeopathy, nutritional medicine, etc. – not one dominated by drug therapy. I stand for more respect for the individual patient and for the natural order.

The Alliance for the Prudent Use of Antibiotics (APUA) (www.tufts.edu/med/apua) was established in the US back in 1981 by Professor Stuart Levy of Tufts Medical School. The main aims of this not-for-profit organisation are to encourage people to take a more responsible approach to the use of these drugs and to promote more informed use of them through the communication of research from different countries around the world. It also seeks to educate people – doctors, patients, students, pharmacists, farmers, manufacturers of animal feed and lay people alike. This kind of international co-operation is a really important step in the right direction.

In the UK, the initiative Antibiotic Action, led by Professor Laura Piddock of the University of Birmingham, was set up to call for the development of new ways of treating bacterial infections. See www.antibiotic-action.com. It is great to see conventional medicine becoming more open to all methods of combating infection.

Chapter 4

Infections to Be Concerned about

Before discussing the most troublesome multidrug-resistant (MDR) infections, it may be helpful to put the present day difficulties in context. Let's go back to 1939 when the Second World War was beginning. If you stood on a nail, scratched yourself accidentally or cut your finger with a bread knife, you would have had to go to your local doctor or casualty department in hospital and have the wound dressed and, if necessary, stitched. Your doctor would have applied a dye such as iodine or gentian violet in the hope of preventing an infection from developing. A few days later, if the wound became inflamed and sore and began to ooze, you would have returned to your doctor or hospital and more than likely the doctor would have cleaned the wound, applied more iodine and a fresh dressing, and asked you to monitor the wound more closely. The doctor may also have taken a swab at this point to confirm a bacterial infection.

Remember that this was the pre-antibiotic era, when there was little in the way of treatment available. Most treatment was aimed at preventing an infection from developing. If preventative measures failed in a case such as the above, a wound infection would most probably take hold. The area where the skin was cut would become covered in pus and eventually form an abscess as the body tried to contain the infection. Medical treatment at this stage would involve draining the abscess and sterilising the whole area to prevent the infection from spreading. If this failed then it was possible the infection would spread deeper into the skin and infect the bloodstream. At this point death was pretty much certain.

So, from a simple scratch or cut you may have lost your life. There were no drugs available to counteract the infection once it had taken hold. If the above case concerned your child, how would you have felt? We are deeply indebted to Dr Fleming, Dr Florey and Dr Chain, and to the drug companies. A parent now is not too concerned if his or her child gets a cut or scratch. Even if the wound becomes infected, he or she knows there are drugs available to help.

Prior to 1940, any form of surgery was risky due to the high death rate from wound infection. Antibiotics essentially removed that risk and all operations and non-surgical procedures became safer, much safer. A whole host of new procedures, such as the angiogram, laparoscopy and colonoscopy, became possible because of antibiotics. Few have died from infections in the Western world since the 1950s.

As we saw in Chapter 1, penicillin changed everything. From its very first use on the policeman Albert Alexander, this drug was revealed as being powerful at combating serious infections. It was truly a living-saving medicine, nothing short of miraculous. It was set to revolutionise medicine and the whole of Western society. Drug companies grew from minnows to giants almost overnight and their close relationship with Western medicine was cemented forever. Patients and doctors felt indebted to these drug companies for eventually bringing penicillin and a host of other antibiotics into commercial production. I grew up in this era when drug companies were viewed in a very positive light.

Prior to 1940, a minor infection often ended in death; today it is regarded as a minor inconvenience. There is a real danger that we may return to pre-antibiotic days if the problem of bacterial resistance is not addressed urgently. The current situation means that superbugs are forming at an alarming pace.

Let's now look at some infections that are currently worrying doctors.

Tuberculosis (TB)

Most of you will be familiar with this infection, but you may not know that it has been making a comeback in recent times. In fact, most people are still of the belief that TB is a thing of the past and is gone forever. Far from it, I fear. Not only is TB on the increase, but it is proving to be more resistant to drugs and as a result is much more difficult to treat.

Today, a third of the world's population is estimated to have the infection, but the majority are symptom free; in other words, they do not know they have the infection as it is dormant in the body – usually in the lungs. According to the World Health Organization, in 2012 no less than 8.6 million people developed symptoms and fell ill with TB. Of these, 1.3 million died in that year (WHO, 2014). It is safe to say then that TB is still a major killer worldwide.

Those with a suppressed immune system are the most vulnerable to contracting the disease. They are 20 to 30 times more likely to die from the infection compared to patients with a normal immune system. In many cases of those who die from TB, the immune system is weakened because of infection with HIV (Human Immunodeficiency Virus). However, immunity can also be weakened by a host of other factors.

Malnutrition is a major factor in the suppression of the immune system, as I witnessed during my trips to southern Africa (Zimbabwe, Botswana, Namibia, Lesotho and South Africa, to be specific). It is, I believe, the main reason why HIV has devastated this region of the world. There are lots of research articles in support of this belief, but one in particular stands out. It showed that administering vitamin A drops to children of mothers who were HIV positive reduced the occurrence of other infections in these children significantly (Coutsoudis *et al.*, 1995). Just one vitamin led to noticeably less illness. How much greater the improvement might have been if these children had access to a nutritious, balanced diet.

Chronic stress is another major factor in weakening the immune system. We all experience acute stress, for instance when speaking in front of an audience, when we pump out adrenalin and our heart rate increases. This is harmless provided it lasts for a short time. If the stress is prolonged over weeks, months or, in some cases, years, the immune system suffers and you become more prone to infection. Bacteria lying dormant in your body can become active, multiply and set up an infection such as TB.

Drugs are another cause of immune system suppression. These include steroids such as prednisolone and hydrocortisone, as well as drugs used to treat auto-immune disorders such as rheumatoid arthritis and drugs such as azathioprine (tradename Imuran) that are used to deliberately dampen the immune response after an organ transplant.

However, the most common reason for a compromised immune system is when the protective layer of bacteria on the body's surfaces has been damaged significantly. This protective layer, called the flora, prevents harmful viruses, bacteria, fungi and parasites from attacking the body.

Many years ago when I was in Africa, I examined a patient with a suspected chest infection in the medical ward. While listening to the front of his chest with my stethoscope, I had to come quite close to him. I asked him to take deep breaths and this caused him to have a fit of coughing that lasted a few minutes. The following day his laboratory report came back and indicated that he had open TB – his sputum or mucus had live bacteria in it, which were capable of infecting others. I was now concerned as I had been close by when he was coughing and had been therefore exposed to these potentially harmful bacteria. However, I did not contract the disease, which suggests that my immune defences were intact; in particular, the bacterial flora that lines my respiratory tract was intact.

Multidrug-Resistant Tuberculosis (MDR–TB)

Due to the inappropriate use of antibiotics we now have a much more serious form of TB – multidrug-resistant tuberculosis (MDR-TB). MDR–TB is a superbug that does not respond to standard treatment. It began to appear in the 1980s and is often accompanied by a HIV infection. In fact, many HIV-positive patients die from TB, usually MDR-TB. Multidrug resistant means that the bacteria have developed resistance to the first-choice group of drugs used to treat TB. Second-choice drugs must be used to treat the disease, but these antibiotics are more expensive and have nastier side effects. In some cases the patient cannot tolerate them.

The patient I mentioned above who had open TB did not have MDR-TB. His TB responded well to a standard drug regimen. If he had infected me, I would have been able to treat myself with the same antibiotic regimen. However, if he had MDR-TB and I contracted it from him, it would have been more difficult to treat and there would have been a greater risk of death.

This is the problem with MDR-TB. It is proving very difficult to treat and the number of cases is growing all the time. In 2012 there were half a million new cases of MDR-TB across the world, according to the World Health Organization (WHO, 2014). This is of great concern to doctors and other healthcare professionals, as they are being exposed more and more to a known killer in their line of work.

TB is intimately related to poor housing, poor sanitation, overcrowding and poor nutrition. These conditions are a breeding ground for the disease and always have been. So, as poverty increases within a country such as the US, the incidence of MDR-TB will increase. As malnutrition increases in certain locations, due to circumstances such as war or natural disaster, we can expect MDR-TB to increase.

Extensively Drug-Resistant Tuberculosis (XDR–TB)

This 'super' superbug is resistant to first-choice and second-choice drugs, making it virtually untreatable. The emergence of such extraordinary levels of resistance has shocked the world and has prompted health authorities to restrict the movement of patients carrying such infections to lower the risk of spread. In 2007, the first case of mandatory quarantine in the US since 1963 made the headlines.

A lawyer from the state of Georgia in the US was planning to get married in Greece. As he was known to have active TB, what was thought to be XDR-TB, he was strongly advised by his doctor not to travel to Greece as he would put fellow passengers on the plane at risk of contracting the infection. Despite this advice, the man went ahead with his wedding plans, so as not to have the hassle of rescheduling, and travelled to Greece. While he was away, the laboratory confirmed he had MDR-TB rather than XDR-TB, but he was still a serious risk to others.

Later that day, his doctor notified the Centers for Disease Control and Prevention (CDC), which in turn called the man and told him not to take a flight back to the US because of the risk of infecting others. Ignoring this advice, he decided to take a flight with his wife to Canada and drive across the border back into the US. He was arrested while trying to enter the US and was placed in mandatory quarantine.

This man was an intelligent professional who understood the risk he posed to his fellow passengers, but yet he went ahead and took a flight to Europe and a flight back to North America. At the time he believed he had XDR-TB, which has a death rate of almost 50 per cent in countries with well-developed medical facilities and a death rate of almost 90 per cent in less well-developed countries. This was not really the best way to start his married life.

His story is far from unique, however. There was the case of the woman who flew from New Delhi to California in

January 2008 knowing that she had been diagnosed with MDR-TB. In California, she presented to a hospital with fever, chest pain and cough. The CDC was notified when her laboratory results showed MDR-TB. This was reported in the journal *Clinical Infectious Diseases* in 2008 (Kaye, 2008).

As you can see from these examples, human behaviour makes it difficult to prevent the spread of highly resistant forms of TB. The truth is that human behaviour is such that one day you or I may be travelling on a plane, train or bus with someone who has XDR-TB, whether they are aware of it or not. We would be at risk of contracting the infection, and there would be little in the way of antibiotics to treat it. The most important thing for you and your family to do to protect yourself against such an eventuality is to eat well and to take a daily probiotic supplement to ensure your bacterial defences are intact (see Chapter 9 on probiotics).

Methicillin-Resistant Staphylococcus Aureus *(MRSA)*

You have already read about MRSA in Chapter 1. This infection is caused by the bacterium *Staphylococcus aureus*, which we carry in our nasal passages. It is a normal constituent of the bacterial flora that colonises our bodies shortly after birth. It is not some alien evil microbe that is out to cause infection and harm us or even kill us. Yet it has the capacity to do precisely that.

When your flora is depleted, this opportunistic organism overgrows and can cause local infections such as boils, carbuncles, abscesses and the like. In other words, this microbe can spread from your nose to cause a localised infection anywhere if your defences are not intact.

As you saw in the case of Albert Alexander, the policeman who was treated with penicillin back in 1940, this bacterium, *Staphylococcus aureus*, was highly susceptible to the antibiotic. However, quite quickly the bug learned how to survive in the presence of penicillin and by the early 1960s the drug companies were trying to devise new antibiotics to overcome the problem of resistance. As mentioned in Chapter 1, by 1962 they had come up with a synthetic version of penicillin, called methicillin, which proved effective at killing this bug.

Twenty years later, *Staphylococcus aureus* found an answer and was showing up as resistant to methicillin as well. Hence, the term MRSA was born. It seemed that clever chemistry in the laboratories of drug companies had been outsmarted by single-celled microbes.

Initially, MRSA infections were limited to hospitals, making hospitals unsafe places to stay in and to work in. Professional staff in some countries used face masks in an attempt to reduce their risk of getting an infection. Others more wisely shored up their body's defences by taking lots of good bacteria as well as eating a very natural diet. The truth is, medical staff and patients alike were very concerned.

One antibiotic was held back as a last resort for severe cases of MRSA. This antibiotic was called vancomycin. By 1996 resistance to this antibiotic showed up in strains of *Staphylococcus aureus*, first in Japan and later in the US. All the major medical authorities in the world appealed for doctors to curtail their use of antibiotics, but unfortunately theses appeals fell on deaf ears.

MRSA then began to show up in the community, and by the early 2000s more cases were being recorded within the community than within hospitals. The superbugs were beginning to rule the roost in the community as well.

As Figure 2 below indicates, community-acquired MRSA has caught up with hospital-acquired MRSA.

Figure 2: Community-acquired MRSA Has Caught Up with Hospital-Acquired MRSA, as seen in the US

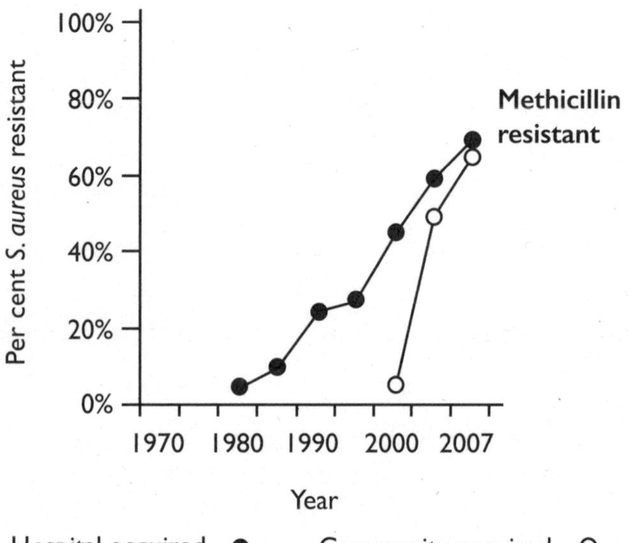

Source: Chambers, H.F. (2001), 'The changing epidemiology of *Staphylococcus aureus*?' *Emerging Infectious Diseases*, vol. 7, no. 2, pp. 178–82. Adapted with permission.

There is another aspect to the MRSA story worthy of mention. MRSA was responsible for localised infections such as abscesses, but not for spreading infections such as cellulitis (inflammation of the skin and just below the skin) or fasciitis (inflammation of the deeper layers of tissue, or fascia). Fasciitis you may recognise by its less pleasant name 'flesh eating infection'.

Prior to 2005, there was no recorded case of *Staphylococcus* causing fasciitis. When I studied microbiology at university and then medicine some years later, I was taught that the *Staphylococcus* species were not capable of spreading horizontally through layers of skin as they lacked certain enzymes to digest human flesh. Clearly, they did not have the capability to cause fasciitis. However, somehow they acquired the genes necessary to cause this flesh eating capability.

In 2005 doctors in the US reported a change in the behaviour of MRSA. Not only was it capable of resisting all the antibiotics we had in our arsenal, but it was also showing it was well capable of inflicting much more serious injury to the human body than was previously thought possible. It was now able to cause deeply invasive infections.

MRSA is not just a superbug but has turned into a flesh-eating monster.

It can be contracted from a door handle, a supermarket trolley handle, a glass, etc. – hence the constant refrain from microbiologists, telling everyone, especially hospital staff, to wash their hands with soap and water. It is vitally important to wash your hands well after you return from a public outing such as a visit to the shops, as you may well contract MRSA anywhere in the community.

Urinary Tract Infections

The biggest bacterial problem facing society today is not MRSA or XDR-TB. You may be familiar with the threats posed by MRSA and resistant TB through newspaper and magazine articles. But there is an even greater bacterial threat to humankind, one that has had less exposure in the media.

It is the infectious disease experts who are ringing the alarm bells loudly this time. They are deeply concerned with the patterns they are witnessing in hospitals across the globe. For the first time they are seeing common infections, such as urinary tract infections, caused by bacteria such as E.coli, becoming impossible to treat. This is due to the fact that these bacteria have extraordinarily high levels of resistance.

It is important to note that E.coli, a normal resident of the gut, is by far the most common cause of urinary tract infections worldwide and always has been. It accounts for about 80 per cent – the other 20 per cent is caused by a variety of other gut residents such as *Pseudomonas*, *Klebsiella* and *Acinetobacter*.

Figure 3 is a laboratory report that I used in my book *Natural Alternatives to Antibiotics* (2003). It is that of a female patient with a urinary tract infection, who I saw back in 1995. As the report shows, the bacterium responsible for this infection is E.coli. When a laboratory isolates a bacterium, it also tests to see which antibiotics would be most effective in treating the infection.

In the report, 'S' means the bacteria are sensitive to the antibiotic and so will die in its presence. 'R' means that the bacteria are resistant to the antibiotic and so the drug will not kill the bacteria.

Look at the number of Rs in the report. This strain of E.coli is resistant to nine antibiotics – it is truly multi-resistant! It is sensitive to only three antibiotics, namely, Netillin, Oflox and Ciproflox. These three antibiotics are rarely used to treat this condition; not only are they quite expensive but they also have nasty side effects.

Figure 3: Lab Report – Urine Analysis of a Patient with a Urinary Tract Infection

Dept of Pathology
Hospital

Date: 03–02–95
Doctor: Dr John McKenna

Sample: Urine

Patient:

Investigation: Culture + Sensitivity

Report: E. coli $> 10^5$
Sensitivity

Amp/Amox	Velocef	Augmentin	Trimeth	Nalidix	Nitro
R	R	R	R	R	R
Gentamycin	Sulpha	Amikacin	Netillin	Oflox	Ciproflox
R	R	R	S	S	S

Source: McKenna, J.E. (2003), *Natural Alternatives to Antibiotics*, Dublin: Newleaf, p. 22.

A report like this, indicating a very high level of resistance, is of course alarming. It is interesting to note that the patient was the wife of a commercial dairy farmer. This pattern of multi-resistance is more common in patients from a farming background because farm animals can carry bacteria that are highly resistant. This is most likely because, as discussed in Chapter 3, antibiotics have been used in animal feed in the past.

A report like this does not augur well for the future. Soon minor infections such as urinary tract infections may require admission to hospital so that antibiotics can be administered intravenously.

In 2006 an article in a microbiology journal showed that

up to 30 per cent of strains of E.coli isolated from urine samples were resistant to the most recently developed antibiotics, namely the quinolones (fluoroquinolones) such ciprofloxacin (Kuntaman *et al.*, 2005). Since then, this percentage has increased. Soon it may become almost impossible to treat one of the most common infections affecting women.

It becomes all the more important for women to protect the vaginal flora. Please see Chapter 7 on vaginal thrush and Chapter 9 on probiotics for ways of doing this. As effective probiotics, I would recommend OptiBac probiotics 'For Every Day' or 'For Every Day Extra Strength'.

It is not just E.coli infections that are of concern. In a New York City hospital in 2000–1 patients were diagnosed with infections caused by other gut bacteria such as *Klebsiella* (Bratu *et al.*, 2005). The laboratory reports do not make for comfortable reading. The strains of *Klebsiella* proved to be resistant to all known antibiotics. In other words, the nightmarish scenario of untreatable infections has arrived.

The other bacteria that are causing concern are *Acinetobacter*. Like *Klebsiella*, they are known to cause infections in patients who have been admitted to hospital for other reasons. They cause thousands of infections every year, most commonly pneumonia, and there are either very few or no antibiotics available to treat these infections. Many of these patients die as the infection cannot be controlled. This is very real evidence of the impasse in which we find ourselves.

If you need to be admitted to hospital for whatever reason, protect yourself with high-dose vitamins and minerals, take drinks containing high doses of vitamin C, such as fresh orange juice with powdered vitamin C added, and eat your own food as generally hospital food is best avoided. Take a probiotic daily and wash your hands well with soap and water.

Be particularly vigilant if you have to have a catheter inserted to drain the bladder as this is a common route for infections of the urinary tract. A number of research articles indicate that about 40 per cent of all hospital-acquired infections are urinary tract infections and most of these have been contracted as a result of having a catheter inserted (Jarvis, 2013).

ANTIBIOTICS CAN CAUSE INFECTIONS

In addition to the problem of bacterial resistance, there are other good reasons to curtail the use of antibiotics. These reasons have mainly to do with the side effects that antibiotics can have. Some of these are commonly known; some are not. Let's start by looking at how antibiotics can cause recurrent infections.

The heading to this section may seem a bit strange as the general perception is that these drugs treat infections. However, there are cases where antibiotics can disturb the body and set up a cycle of recurrent infection. The following case history is a good example:

Case History: Jonathan

Jonathan was seven years old in 1993. Between February 1993 and May 1994 he was treated for ear infections and chest infections. Table 1 below provides a list of the antibiotics he was prescribed.

Table 1: Antibiotics Prescribed to Jonathan, 1993–4

Date of Prescription	*Name of Medication*
11 February 1993	Distaclor
18 March 1993	Distaclor
17 May 1993	Augmentin
30 May 1993	Septrin
6 June 1993	Distaclor
5 July 1993	Septrin
10 July 1993	Septrin
25 August 1993	Augmentin
1 September 1993	Augmentin
10 September 1993	Augmentin
27 September 1993	Distaclor
23 December 1993	Augmentin
4 January 1994	Septrin
24 February 1994	Erythroped
27 February 1994	Distaclor
8 May 1994	Distaclor

This child was first treated for an ear infection on 11 Feb 1993. By May of the following year he had a total of 16 courses of antibiotics for recurrent ear infections as well as an occasional chest infection. Most of these antibiotics are broad spectrum, which means they kill not only bad bacteria but can also disturb the natural protective bacteria in the body – the good bacteria – to the point of setting up a cycle of recurrent infection.

Lots of scientific research exists to validate that early treatment of common childhood infections can cause long-term problems. As far back as 1987 an article appeared in the journal *Pediatric Infectious Diseases* that illustrated just that in the case of treatment of a streptococcal sore throat. The article states that if an antibiotic is used in the first two days of the infection then the infection is two to eight times more likely to recur. In other words, if you treat the infection sooner rather than later you are much more likely to get another sore throat. The article concludes that it is better to withhold treatment for a minimum of 48 hours (Pichichero *et al.*, 1987). Even earlier than that, in 1974, an article in the journal *Archives of Otolaryngology* looked at the treatment of middle ear infections in young children and came to a similar conclusion. This article also suggested holding off antibiotic treatment for as long as possible to avoid the risk of setting up a cycle of recurrence (Diamant and Diamant *et al.*, 1974). So we have known for almost forty years that being too fast to prescribe an antibiotic can cause ongoing problems.

One of doctors who warned us of the dangers of treating ear infections too early was an international authority on ear disease in young children. His name was Dr Erdem Cantekin and he was director of a research centre in the University of Pittsburgh, Pennsylvania in the US. He was co-investigator in a five-year study to investigate the effectiveness of certain antibiotics in the treatment of middle ear infections in children.

The first phase of this study was funded by the US National Institutes of Health, which is to say it was funded by public money; the second phase was funded by the pharmaceutical company that manufactured one of the antibiotics being tested – amoxycillin. In other words, the funder of the project had a very strong vested interest in the outcome. This was a scenario fraught with difficulty (Rennie, 1991).

Dr Cantekin's results showed that amoxycillin was ineffective, which was not exactly what the pharmaceutical company wanted to hear. He was then promptly fired as director of the research centre and forbidden to publish his results. Despite this, Dr Cantekin tried to publish his findings, but the *New England Journal of Medicine* and the *Journal of the American Medical Association* both refused to publish them (Rennie, 1991).

Meanwhile, another doctor who had accepted money from pharmaceutical companies for lecture fees and travel expenses, as well as research, was appointed as director of the research centre (Rennie, 1991). He published results in the *New England Journal of Medicine* in 1987 showing that amoxycillin was effective (Mandel *et al.*, 1987). Doctors across the world prescribed amoxycillin thinking this was the case until four years later the *Journal of the American Medical Association* agreed to publish Dr Cantekin's data. His data showed that amoxycillin was ineffective in treating middle ear infections in children; two other antibiotics, Distaclor (the tradename for cefaclor, used in the case study above) and Pediazole (a combination of erythromycin and sulfisoxazole), were also ineffective (Cantekin *et al.*, 1991).

Dr Cantekin's data also showed that children treated with amoxycillin were two to six times more likely to suffer a recurrence of the infection. The conclusion of his findings was that these antibiotics were not effective in either acute or chronic cases of middle ear infection.

Most medical research is funded by pharmaceutical

companies, as are most medical journals, which rely on advertising revenue. Conflicts of interest can result in the truth being suppressed for financial gain. This is clearly not acceptable. Nor is it acceptable that doctors who challenge the system or the old boy network be punished and ostracised.

Dr Cantekin's research data is very revealing. It shows what we have known for many years now: not only should we delay treatment of sore throats and ear infections but we should be selective about which antibiotic to use.

As mentioned, the best means of deciding which antibiotic to choose is to send a swab to the laboratory and ask for a culture and sensitivity profile. In this way you can confirm whether it is a bacterial infection you are dealing with in the first place. This should really be done for all infections to curb the misuse of antibiotics. If there are bacteria on the swab, these are cultured in the lab and identified. They are cultured in the presence of certain antibiotics to see which ones inhibit their growth.

So antibiotics used to treat a throat or ear infection can cause recurrence of that same infection, if the drug is used too soon. But it appears from other research that antibiotics can also cause other infections. Research published in the *Archives of Diseases in Childhood* in 1991 showed that antibiotics can cause changes in the flora of the urethra of children (Lidefelt, K.J. *et al.*). The article suggests that this may predispose children to a urinary tract infection.

The results of this research are not really surprising as antibiotics are known to alter the bacterial population of the body in adults and cause infections such as vaginal thrush and intestinal candidiasis. Clearly the same phenomenon occurs in children. To date, I have found nine other articles confirming this finding.

Since antibiotics have the potential to cause recurrent infections or predispose children and adults to other infections, such as urinary tract infections, it is wiser to avoid them as best you can. After all, most infections in children are viral and do not require an antibiotic. It is wiser to assume every infection is viral until proven otherwise. Avoid an antibiotic until a swab has been collected and the results show conclusively the infection is bacterial.

If you or your child has an infection, the best course of action in the first instance is to use antiviral measures: large doses of vitamin C, zinc and, in some cases, vitamin A, either as betacarotene or as retinol, as well as a range of homeopathic remedies. For further information, refer to my book *Natural Alternatives to Antibiotics* (2003), which covers a range of alternatives that you can use instead of these drugs.

Today it is accepted wisdom that antibiotics can cause infections if they are used too soon in the course of an infection. Having said that, many doctors and patients turn to antibiotics early to avoid the hassle of taking a swab or 'to be on the safe side'. If you have an infection, you are putting your health at risk by not delaying treatment. You are also adding to the already very serious problem of bacterial resistance. It is much wiser to wait. Use an antibiotic as a last resort and only when the infection has been proven to be bacterial and not due to a virus or fungus.

On the subject of fungal/yeast infections, Chapter 5 takes a look at the damage that antibiotics can cause in the gut.

Chapter 5

How Antibiotics Affect the Gut

Most drugs have side effects that affect the gut but these are usually mild, nausea being the most common. You are normally aware of the negative effects quite soon after taking the drug. Antibiotics are a bit different as in some patients the effects are initially unseen, but they can have a detrimental impact on your health for years afterwards. Here is an example of what I mean.

Case History: Jane, Itchy Skin Rash

Jane was four years old when her mother brought her to see me because of a troubling skin rash. This rash had been there for months, despite treatment from her doctor with steroid creams and antibiotic creams. It was affecting the perineum, which is the area between the vagina and anus, and was very itchy. From Jane's medical history the only thing noteworthy was that she'd had recurrent ear infections in the first year of life and had been treated with antibiotics. Prior to the onset of the rash, she had taken another antibiotic for a chest infection.

I suspected she had the skin manifestations of dysbiosis, a disturbance in the balance of the bacterial population, probably as a consequence of the numerous courses of antibiotics. I treated her for intestinal dysbiosis and asked her mother to coat the affected area of skin with a probiotic solution. Within two months Jane's skin problem was resolved. She also had more energy and was generally a lot better. She continued with treatment over the following six months and has remained in good health since.

Jane's case history is an example of the unseen damage that a course of antibiotics can do. Such damage is not often suspected or recognised in medical practice as doctors are not taught about the bacterial flora of the body. I spent nine years in university studying microbiology and then medicine, but in all that time I didn't learn anything about the normal bacterial population of the body. I learned lots about bad bacteria but nothing about good bacteria. I learned a lot about antibiotics but nothing about probiotics. This is still the case for medical students today. It was only when I did a postgraduate course that I learned about things such as dysbiosis and probiotic supplements. There is clearly a massive gap in medical education as the bacterial flora of the body plays a pivotal role in human health.

Good bacteria line the digestive tract, the respiratory tract, the skin and the vagina in females. For every human cell in your body, you have nine bacterial cells. Therefore, 90 per cent of your body is made up of good bacteria. You are coated in trillions of bacteria. If this population of bacteria gets disturbed then there are usually manifestations in the form of ringworm or athlete's foot on the skin, vaginal thrush, dysbiosis in the gut, or chronic bronchitis or persistent cough in the respiratory tract.

Let's now look at the effects of antibiotics on the gut.

Antibiotics Can Cause Intestinal Dysbiosis

When the ecological balance of good bacteria in the digestive tract is disturbed we call this dysbiosis. It implies that some of the good bacteria have been killed off by drugs such as antibiotics and that bad bacteria have taken their place. I do not like the terms 'good bacteria' and 'bad bacteria', as they are too simplistic. The truth is that when you kill some of the normal bacterial flora in the gut, other members of the bacterial flora overgrow and take the place of the dead ones. It is really the overgrowth of microbes such as *Candida* that caused the problem. Russian biologist Dr Élie Metchnikoff was the first person to use the term 'dysbiosis' at the beginning of the twentieth century. Dr Metchnikoff succeeded Louis Pasteur as director of the famous Pasteur Institute in Paris. He suggested that many digestive diseases resulted from dysbiosis. He maintained that problems such as peptic ulcers, diarrhoea, constipation and even cancer of the colon were all the end result of disturbances in the ecology of the bacterial flora in the gut. He won a Nobel Prize for his work on the bacterial flora of the gut in 1908.

Dysbiosis means there is an abnormal bacterial population in the digestive tract: an overgrowth of a microbe such as *Staphylococcus* or *Candida.* The symptoms associated with dysbiosis are for the most part not obviously related to the gut. They include constant tiredness, a 'spacey' feeling, moodiness, headache, erratic vision, vaginal discharge, skin itchiness, skin rashes, poor memory and poor concentration. These can be interpreted as quite general symptoms and are often not connected to previous exposure to antibiotics, mainly because it can take months or years after a course of antibiotics is taken for the symptoms to develop. Some patients have no gut symptoms, or are not aware of them until asked specifically about them. Usually the more general symptoms provoke people to come for help, as these can ruin one's quality of life.

Most doctors do not recognise dysbiosis, despite the fact that it is a well-known and real disorder. So, if you present to your doctor with low mood and low energy, he or she will typically check your iron status, vitamin B12 level, thyroid function, etc. If these tests come back normal you may be referred to a psychiatrist or just dismissed.

The following case histories illustrate the symptoms of and treatments for dysbiosis.

Case History: Kathy, Depression

Kathy was in her late twenties when she came to see if I could help her get to the bottom of her mood swings and irritability. When I questioned her it became clear that her mood was more consistently low than up and down, and her energy levels were also very low. Her bouts of depressed mood were interfering with her home life and work life, as her husband was finding her quite grumpy and her work colleagues were finding her edgy and irritable.

In her teens and early twenties she had been on an extended course of the antibiotic tetracycline for acne and it was since then that she noticed a dip in energy and mood. Tests carried out by her GP, including thyroid function tests, were all normal. My tests revealed significant dysbiosis and so I treated her with changes to her diet and nutritional supplements.

I also asked her to check her thyroid function by measuring her basal body temperature, the lowest temperature attained by the body during rest, which she obtained by taking her body temperature upon waking on three consecutive mornings. Her results were almost within the normal range of 36.4–36.7 degrees Celsius and so I could rule out subclinical hypothyroidism, which means low thyroid function but not low enough to show up on a blood test.

After the first two weeks of treatment her husband and workmates had positive things to say about her mood, but it took another six weeks for her to begin to feel her energy normalising. She had to stay on treatment for another 18 months because her condition was long standing. Dysbiosis is also quite slow to heal, mainly because you are dealing with trillions of bacteria.

The next case history deals with the very prevalent issue of chronic tiredness. Fatigue is the most common complaint presented in doctors' surgeries. It is generally caused by low iron levels or anaemia, low thyroid function or low blood glucose. But it can be also due to a disturbance in the gut flora, as the following case indicates.

Case History: John, Constant Tiredness

John was 30 years old and in the prime of his life. He had a very successful career as a lawyer and was happily married. He found that when he played five-a-side soccer he did not have the same stamina as before. He also woke each morning groggy and exhausted as if he had not slept at all. His energy picked up after breakfast but could dip a bit mid-morning and dip significantly in the afternoon to the point where he found it hard to keep his eyes open. After supper, he often dozed off in the chair while watching television. He also had a lot of abdominal bloating after eating and his tongue had a constant white coating.

On examining him, I saw that his tongue had indeed got a white coating and he had mild distension of the abdomen. He also showed evidence of athlete's foot and a fungal infection on some of his toenails.

Tests showed that he had dysbiosis and a deficiency in some of the B vitamins and in magnesium. Treatment involved changes to his diet and nutritional supplements, including a good multi-mineral/vitamin supplement. He also used Lamisil for the athlete's foot.

As soon as he began the multi-vitamin/mineral supplement, his energy improved, but it took much longer to resolve the other problems.

This case history illustrates one of the consequences of dysbiosis – vitamin deficiencies. The good bacteria that line your digestive tract manufacture a number of B vitamins, which are used by the body. When these good bacteria are killed off you can easily become deficient in one or more of the B vitamins. In addition, bad microbes that overgrow can steal the B vitamins you may consume in food, rendering you even more deficient. So if you suspect you have dysbiosis, take a good multi-vitamin/mineral supplement.

Case History: Brian, Feeling 'Spacey'

Brian was 28 years old and was worried that there was something seriously wrong with him. During the last year he had been having difficulty concentrating and his short-term memory was letting him down to the point where he had to write everything down. Most disturbing for him was the constant feeling of being 'a bit spacey', where his thinking was muddled. He had never experienced such symptoms before and was convinced that he had a brain tumour or some other malevolent disease. Extensive testing by his GP *and consultant neurologist revealed nothing abnormal. This was when he came to see if I could shed any light on his problem.*

Since feeling spacey is a cardinal symptom of dysbiosis, which many conventional doctors fail to diagnose, and since Brian had other symptoms and signs suggestive of dysbiosis, I had a few tests done. The results did indeed indicate dysbiosis; in fact, Brian had what is described as fermentation dysbiosis, which means that there is an overgrowth of yeast combined with the dysbiosis. The fermentation was obviously making his symptoms worse than if he had dysbiosis alone. Fermentation meant that any sugars or starches that he ate were being partly converted to ethanol (alcohol). And of course, as we know, alcohol interferes with your brain function.

I treated him over the coming months and slowly but surely he started to mend. After six months of treatment he found that his memory had improved and he did not need to write notes for himself any longer.

How does dysbiosis cause symptoms such as Brian's? Both fermentation and dysbiosis result in toxic chemicals being produced, such as ethanol, methanol and propanol. These chemicals not only interfere with liver function *but can also impair brain function*; hence, the lack of concentration, memory loss, and the strange spacey feeling and cloudy thinking. These are really symptoms of auto-intoxication, where the toxin is generated by the body and not ingested in food or water, or inhaled. In other words, the body is slowly poisoning itself because of a disturbance in the bacterial flora. For this reason alone, I cannot over-emphasise the importance of good bacteria to your health.

Case History: Paula, Headaches

Paula was in her early twenties and was suffering from severe headaches. She had a sensation of pressure inside her head and at the back of her eyes. On occasion, the headaches were accompanied by feelings of weakness and nausea. She had them about two or three times a week and, although painkillers did help, she was worried about having to take painkillers so frequently. The headaches were also interfering with her quality of life. Her GP carried out numerous tests and they were all negative. No cause could be found for her headaches. This was when she decided to come to me for help.

When I examined her I found a lot of tension in her neck muscles and in her upper back. I was not sure if this was contributing to her headaches, so I suggested she have some treatment to release the tension while we were waiting for the results of her tests. The stool test revealed a moderate overgrowth of Candida albicans. *I explained to her that dysbiosis can cause toxic headaches and since the description of her headaches sounded more like they had to do with toxicity than tension, I suggested that she treat the dysbiosis for a period of two months initially to see if this alleviated the headaches. It was three months later when I saw her again. I got a big hug from her as she was ecstatic that her headaches had improved enormously.*

They only recurred when she broke her diet or stopped the supplements.

Paula's case shows how toxic the body can become when you have dysbiosis. I am sure you've seen mould growing on food that has been exposed to air for too long. Moulds or fungi have roots that extend down into the food. When fungi grow on the intestinal wall, as can happen with dysbiosis, they also develop roots called mycelia, which penetrate through the gut wall into the bloodstream. All the fungal waste products are then dumped directly into the bloodstream. Dysbiosis can also facilitate the development of fermentation (maldigestion of carbohydrates) and putrefaction (maldigestion of protein), both of which can result in the production of very toxic chemicals that can also cause severe headaches.

If you suffer from recurrent headaches, you should find the underlying cause rather than taking painkillers, as these only add to the toxic load. Get checked out to see if you have dysbiosis.

Dysbiosis can also present with abdominal symptoms only, as the next case history shows.

Case History: David, Diarrhoea

David ran a grocery shop and had to work long hours. He was 33 years old and was married with two children. He had been suffering from diarrhoea on and off for a few years. His stool was poorly formed most of the time – 'a mushy mess', to use his own words – and on occasion appeared very loose and had mucus in it. In addition, he complained of abdominal bloating and lots of wind, which caused him a lot of embarrassment. As he also had a lot of perianal itching, he was convinced he had worms. After two doses of worm medicine from his doctor, he found that the symptoms were still there and so he came to see me. On questioning David, I found out that he had been on antibiotics for about three years. He had also received lots of antibiotics and steroids as a child for severe asthma.

A stool test revealed two species of fungi as well as mucus. I began David's treatment with colonic irrigation because of the high level of mucus in his stool sample. A probiotic solution was instilled into the colon after each treatment. I then treated the dysbiosis and included flaxseed oil capsules to help bulk the stool, leading to better bowel clearance.

After two months of treatment, David had a normal, well-formed stool and no bloating or wind for the first time in years. His digestive tract was on the road to recovery.

David's case illustrates the abdominal symptoms associated with dysbiosis. A loose stool is common with this condition, although some people have constipation and others alternate between constipation and diarrhoea. Most people do not seek help when the consistency of the stool varies like this; they often dismiss it as due to dietary changes and so on. Bloating and wind are also common symptoms of dysbiosis, but again they are usually not seen as severe enough to warrant medical attention. It is only when a symptom becomes severe that people seek assistance.

Other abdominal symptoms include abdominal cramps and sometimes abdominal pain. Less common symptoms include heartburn and indigestion; these are usually caused by intolerance to gluten (wheat, rye and barley).

So, in summary, dysbiosis can have a few abdominal symptoms but a lot of general or non-abdominal symptoms.

HOW IS DYSBIOSIS DIAGNOSED?

The simplest method is to base the diagnosis on the patient's medical history and a clinical examination. This is usually sufficient for any experienced practitioner. To confirm one's clinical suspicions, it is possible to order various tests. A stool analysis is one that I found reliable when I first began to practise. However, a negative stool test result does not necessarily mean that the person's flora is normal; the piece of stool selected for the test may not have contained abnormal levels of certain strains of bacteria or fungi or yeast. Also, the stool sample may not have been handled correctly in transit to the laboratory or there may have been an error made in the laboratory. If a test comes back negative I have to trust what the patient tells me and what I find from clinical examination. Results are always of secondary importance and should only be used to confirm one's clinical suspicions.

It's also possible to diagnose dysbiosis by means of blood tests. One of these blood tests is called a gut fermentation test. In this test the patient is given glucose to swallow and one hour later a blood sample is taken. Various alcohols are measured in the sample. Figure 4 shows an example of the results of a gut fermentation test.

Figure 4: Lab Report – Gut Fermentation Test Results

GUT FERMENTATION PROFILE

Note: A glucose load was given one hour before sampling.

	Result (μmol/l)	Reference Range:
Alcohols: In most cases, it is only the ethanol result that is affected by the glucose load.		
Ethanol	63	less than 22
	0.3 mg/dl	less than 0.1
Methanol	1	less than 2.5
2-propanol	1.3	less than 1.0
1-propanol	1.7	less than 0.5
2-methyl-2-propanol	0.4	less than 0.3
2-methyl-1-propanol	N.D.	less than 0.3
2-butanol	1.3	less than 2.3
1-butanol	2.4	less than 1.2
2-methyl-2-butanol	N.D.	less than 0.5
2-methyl-1-butanol	0.5	less than 0.3
2-ethyl-1-butanol	0.3	less than 1.0
2, 3-butylene glycol	2	less than 2.5
Short chain fatty acids and related substances:		
Acetate	79	52–85
Propionate	37	10–56
Butyrate	13	7–33
Succinate	11	1–31
Valerate	11	2–19

N.D. – Not detected

Comment: All results assume 24 hours without alcohol ingestion and 3 to 12 hours fasting.

As you can see in this result, some of the alcohol levels are above the reference range. A raised ethanol value suggests a yeast overgrowth. In Figure 4 the ethanol value is 63 micromoles per litre – it should be less than 22 micromoles per litre. Some of the other alcohols are also shown to be above the reference levels: 2-propanol, 1-propanol, etc. This suggests the presence of dysbiosis in this person.

Another blood test that can be used to diagnose dysbiosis is a D-lactate test.

The presence of D-lactate in humans suggests that there is an overgrowth of colonic bacteria. High D-lactate levels are toxic to the body as they make the body's fluids too acidic, a condition called a metabolic acidosis. This test is much simpler to carry out.

HOW IS DYSBIOSIS TREATED?

There is a lot of controversy about how best to treat this condition. However, while books may vary quite a lot on the specific details of treatment, the principles of treatment are basically the same across the board. The four main principles are as follows:

1. Do not feed the bad bacteria – in other words, diet is important
2. Kill off the bacterial/yeast/fungal overgrowth – there are a range of antimicrobials on the market
3. Replace the good bacteria – probiotics
4. Replace lost nutrients – a good multi-vitamin/mineral supplement

Let's examine each of these aspects of treatment in detail.

Diet

Some books suggest eliminating sugary foods only, while other books suggest eliminating yeast foods as well. Some practitioners restrict cereals, while others restrict all carbohydrates. The truth is that the correct diet for you will depend on your body's response. A gut fermentation test will show if it is necessary to restrict all carbohydrates – if the ethanol value is high then all carbohydrates should be restricted; a leaky gut test will indicate if it is necessary to restrict gluten cereals (see my book *Hard to Stomach* (2002)).

If it is not possible to do any of the tests mentioned above then I would suggest following stage one of the gut fermentation diet outlined in Table 2 for one month and then adding in the foods mentioned in stage two when you have tested them one at a time. If any of these foods cause a negative reaction, such as drop in energy, abdominal bloating or indigestion, then omit it from your diet.

Table 2: The Gut Fermentation Control Diet

This diet is in four stages, each lasting one month. Do not proceed to the next stage until you have been examined and re-tested. Do not stop any medication that you have been prescribed in stage 1 until you've seen your practitioner.

Stage 1: eat only the following foods:

Meat	all sorts, including beef, venison and lamb. No processed meats
Poultry	including chicken, turkey and guinea fowl. Try to use organic poultry
Fish	not in batter

Vegetables	all types except mushrooms, sweetcorn and peas. Best steamed stir fried or baked
Fruit	pawpaw (papaya), rhubarb and grapefruit are the best to use
Salads	
Soups	do not add potato as it is too starchy
Cheese	excluding all soft cheese, e.g. cottage and cream cheeses

---

Stage 2: try each of these foods one at a time for two to three days. If you tolerate it well, continue with it; if you can't tolerate it yet, delete it from your diet for another month.

Bread	must be yeast-free
Rice cakes	
Rye	Ryvita and yeast-free bread, e.g. sourdough rye bread
Fruits	try each new fruit one at a time
Oats	
Dairy produce	
Rice	best forms are basmati rice or brown rice

Stage 3: in this stage you can add in yeast-containing foods, such as:

Breads	including rolls, croissants, pastry and doughnuts
Pickles	
Vinegar	
Alcohol	wine and beer are fermented and so contain yeast
Dried fruits	raisins, currants, sultanas, prunes, dates

Malted cereals and malted drinks

Mushrooms (they are a fungus)

Gravy mixes
Spreads
Yoghurts
Yeast, e.g. brewer's yeast

Stage 4: in this stage you can add in sugar-containing foods, such as:
Jams/marmalades
Breakfast cereals
Desserts, cakes and biscuits
Soft drinks
Sugar, confectionery and chocolate

Source: McKenna, J.E. (2002), *Hard to Stomach: Real Solutions to Your Digestive Problems*, Dublin: Newleaf, p. 39.

It is very important to try each of the new foods in stage two one at a time and for two days at least so that you get a good idea whether the food is causing an aggravation or not. Do not go to stage three until you have followed stage two for three months or until a repeat test indicates that the flora has normalised. So your diet should end up consisting of all the foods allowed in stage one of the fermentation diet and any foods from stage two that do not cause adverse reactions. The best form of starch to use and also the food least likely to cause an adverse reaction is rice – brown or basmati being preferable.

Antimicrobials

Killing off the bad microbes is an important aspect of treatment. Some people believe that it is possible to treat dysbiosis with diet alone; others believe that diet plus taking a probiotic is all that is necessary. However, the single most important aspect of treatment is to rid the body of the overgrowth of bad microbes.

There are two main antimicrobials used by practitioners of natural medicine. The more potent of the two is grapefruit seed extract. Its advantage is that it is not just antifungal but is also antiparasitic and antiviral. It can therefore be used regardless of the type of dysbiosis and the type of parasite. Its main disadvantage is its very bitter taste. A way around this is to take it in capsule form but make sure to drink lots of water with it.

The second most popular antimicrobial to treat dysbiosis is a substance called caprylic acid, a fatty acid that occurs naturally in coconuts. It has proven antifungal properties and the calcium and magnesium salts – calcium caprylate and magnesium caprylate – survive the digestive juices of the upper gut and so can act on the large intestine. Since most forms of dysbiosis involve the overgrowth of yeast or fungi within the colon, caprylic acid is a very useful medicine for this condition.

It is very important to remember that killing off microbial overgrowths in the gut can make you more toxic. Millions of dead cells can rupture and the waste products can enter the body and make you feel worse. Therefore, it is important not to get constipated, to allow for the disposal of these products. It is also a good idea to have colonic irrigation done, which would guarantee the elimination of dead organisms.

Probiotics

The ecological balance of the intestinal flora can be restored to normal by taking lots of good bacteria, either in live yoghurt or in a probiotic supplement. Probiotic supplements contain a number of species of good bacteria, such as *Lactobacillus acidophilus* and *Bifidobacterium bifidum*. Probiotic supplements are usually prepared as freeze-dried capsules or powder. When water is added to the capsule or powder, the bacteria are regenerated and can multiply. It is important to keep the probiotic supplement cool and dry so it is best stored in a fridge. The decision on dosage will depend on the degree of disturbance to the flora and so is best left to your practitioner.

Health food shops have lots of probiotic supplements to choose from. Some of these are dairy-free, for people sensitive to cow's milk. When bacteria are regenerated and multiply, they need food to survive. The more modern supplements may contain a *prebiotic*, which is a compound that acts as food for the good bacteria. There are two main prebiotics in use, inulin and fructooligosaccharides (FOS). Both of these prebiotics are excellent substances but can also cause a lot of flatulence, so do not use if you have a lot of wind already or if you have colitis or Crohn's disease.

Live yoghurt or a probiotic supplement should be used by everyone because every day we lose good bacteria from the digestive tract in the same way that we shed skin cells. It is best to take these good bacteria first thing in the morning, as the stomach is much less acidic at that time and so more of the bacteria will survive. It is also best to take the good bacteria on an empty stomach and wait a while before having tea or coffee as a hot drink may kill them.

A Multi-vitamin/Mineral Supplement

One of the reasons why good bacteria are important is that they manufacture some of the B vitamins that help your body to function effectively. Loss of these good bacteria means loss of essential vitamins. When bad bacteria and fungi overgrow, they rob you of some micronutrients. Therefore, a good multi-vitamin/mineral supplement becomes important. Your health practitioner will advise you on a suitable one for you.

ANTIBIOTICS CAN CAUSE COLITIS

Apart from dysbiosis, antibiotics can cause another disorder of the gut – inflammation of the colon (colitis). The link between taking an antibiotic and colitis is outlined in the case histories below. These case histories show the range of bad effects that antibiotics can have on the colon. Colitis can be a very serious complication, as Frederick's case history below illustrates.

Case History: Catherine

Catherine was put on antibiotics to treat her acne. She was prescribed Minocycline for six months initially. After three months of being on the drug, she developed diarrhoea with blood and mucus in the stool. Her doctor suggested that she stop taking the antibiotic and use another antibiotic instead. After one month on this new antibiotic, she developed diarrhoea with blood and mucus again. So her doctor decided to use hormonal treatment instead. Upon stopping the course of antibiotics on both occasions, the diarrhoea resolved and did not recur.

Case History: Marianne

Marianne had a bad chest infection and was put on two broad-spectrum antibiotics to treat it. After three days of being on this treatment, she developed severe diarrhoea with a lot of mucus in the stool and a small amount of blood. Stopping the antibiotics did not result in a lessening of the diarrhoea, so she was taken into hospital. Here, her chest infection was treated via intravenous antibiotics and a colonoscopy was carried out. She was diagnosed as having colitis, which resolved with the use of steroids.

Case History: Frederick

Frederick was on a course of antibiotics for recurrent streptococcal sore throats. He had taken many courses of antibiotics for this problem in the past. Like Marianne, he developed colitis as a consequence of taking the antibiotic. The colitis got worse on stopping the drug, and he had to be admitted to hospital as he was passing a lot of blood in his stool. In hospital he was diagnosed with a Clostridium difficile *(C.diff) infection. Despite the use of further antibiotics to treat this pathogen, his condition got progressively worse and he died.*

The use of an antibiotic to treat an infection may seem like an automatic and safe thing to do, but in a small number of cases it can have fatal consequences. The onset of diarrhoea upon starting a course of antibiotics must always be taken very seriously. Fortunately, most cases will resolve quickly if the antibiotic is stopped. If the diarrhoea persists, it is important that the patient is sent to hospital as they may have a more serious form of colitis, such as that suffered by Frederick above.

Frederick was diagnosed with C.diff, an infection that is becoming more frequent in hospitals and care centres for the elderly. This bacterium is a normal constituent of the population of bacteria in the gut. In other words, we all have C.diff in our bodies, and at low levels it is of benefit to us. When we take a course of antibiotics we can damage the normal bacterial population and C.diff then takes advantage and begins to overgrow. If it overgrows sufficiently it can dominate the population of bacteria in the large intestine. It then produces a nasty toxin that can irritate the wall of the large intestine and cause colitis.

The symptoms of colitis caused by C.diff are watery diarrhoea, fever, loss of appetite, abdominal pain or abdominal cramps. The diarrhoea may or may not have blood or mucus in it. However, because of the severity of C. diff infections, it is wiser to investigate all cases of antibiotic-associated diarrhoea.

C.diff infections are a great worry for hospital staff as the bacterium is easily spread from one patient to another. Therefore, patients with a diagnosed C.diff infection need to be isolated and all staff need to use gloves and gowns when treating such patients. C.diff can form spores, which can live on sinks, worktops, counters and other inert surfaces. These spores are very hard to get rid of as they are not killed by the hand sanitisers and disinfectants we have become accustomed to using. The spores of this microbe can only be killed by strong bleaches. To effectively prevent infection, one has to bleach all surfaces regularly.

The people most susceptible to C.diff infection are the elderly, patients with impaired immunity such as those on immunosuppressive drugs, those who are HIV positive, patients on more than one antibiotic and patients on long-term antibiotic treatment. It is a very serious infection – the Centers for Disease Prevention and Control (CDC) in the US reports that C.diff is responsible for 14,000 deaths per year (Clifford McDonald *et al.*).

Treatment can prove difficult, especially if the bacterium is highly resistant, as many hospitals are finding to be the case. Metronidazole and vancomycin are used to treat it, but even after treatment 30 per cent of patients will get a recurrence of colitis and some, like Frederick, may die.

ANTIBIOTICS CAN CAUSE CANDIDIASIS

Candidiasis refers to an overgrowth of *Candida albicans*, a yeast-like microbe that is a normal constituent of the gut flora. It overgrows when the bacterial population gets disturbed by stress or drugs such as antibiotics, especially broad-spectrum antibiotics like amoxycillin, ampicillin, tetracycline and metronidazole.

What I find amusing is that many doctors deny the existence of candidiasis, despite the fact that it is discussed in established and respected medical textbooks such as *Davidson's Principles and Practices of Medicine* and pharmaceutical textbooks such as *The British National Formulary*. Because doctors are causing this disorder with their misuse of antibiotics, I suppose they would prefer to ignore the consequences of their actions.

Candidiasis usually begins in the gut but later on can cause symptoms in many other organs of the body. In the gut it can cause a loose stool, indigestion, flatulence and bloating. As the flora gets further disturbed, food intolerances develop and you can begin to react to foods that you tolerated before. Then cravings for certain foods develop. Later, you can go on to develop systemic symptoms such as tiredness, feeling spacey and being irritable or moody. My own case history is a good example.

Case History: John

When I was at medical school, my diet was awful as there was never sufficient time to eat. I lived on hamburgers and chips, hotdogs, sandwiches, etc. I ate a diet of convenience; the more convenient the food, the more I ate of it. At that time I was suffering from what I thought were streptococcal sore throats and had been given numerous courses of antibiotics. I then developed a skin rash and was put on tetracycline, a broad-spectrum antibiotic, for three months. Since there was little improvement, I went to my professor of medicine who put me on another antibiotic for six months. The rash was getting worse and I was feeling very tired all of the time. So I decided to consult a naturopath, who diagnosed candidiasis and put me on treatment. Almost immediately, I started to feel better, and within a few months I was well improved.

This experience showed me the deficiencies in conventional medical thinking and made me acutely aware of the benefits of natural medicine. It was hard to accept that doctors could not help me. They were my teachers and I looked up to them. It was equally hard to accept that a non-medic was so accurate and so helpful. Natural medicine had helped where conventional medicine had not only failed but had made my condition worse. It was a lesson that opened my eyes, since it had such a profound effect on my own personal life. It wasn't someone else's story or some bit of theory. Rather, it was something I could not deny. It sparked my interest in the misuse of antibiotics.

ANTIBIOTICS CAN PREDISPOSE YOU TO PARASITES SUCH AS GIARDIASIS

As the use of antibiotics has increased over the years, the incidence of parasitic infections has also increased. The two are directly related; antibiotics make you more susceptible to parasitic invasion.

When antibiotics damage the bacterial population in your gut, they not only predispose you to the overgrowth of opportunistic microbes such as *Candida albicans* and C.diff but can also make you more vulnerable to parasites in water and food, and in the external world in general. One common parasite is *Giardia lamblia* and the infection it causes is called giardiasis.

Giardia lamblia is found all over the world but is much more common in areas where sanitation is poor or the water supply is unsafe. It is spread mostly through water, but it is also spread through food and person-to-person contact. It causes tummy upset and diarrhoea as well as general malaise. Usually it clears up in a few weeks without treatment, but in some cases it can persist.

Source of Infection

The most common mode of transmission by far is through contaminated water. *Giardia* has been found in ponds, lakes, rivers and streams in many countries worldwide. It has also been found in tap water, and in wells that supply homes and offices. Despite the presence of chlorine, it has been discovered in swimming pools and in spas.

The reason for it being found in so many water sources has to do with the contamination of these supplies with faeces from animals or humans, or with waste water from farms or homes. The contamination of swimming pools and spas has to do with children wearing nappies in the water or adults/children with diarrhoea entering the water.

Giardia finds its way into food through food handlers not washing their hands before preparing the food. Foods can also become contaminated when raw foods such as salads are washed with water infected with *Giardia*. High temperatures will kill this parasite, so cooked food is perfectly safe. If travelling to areas of the world with poor sanitation, always eat cooked food and drink only bottled water.

This parasite can be transmitted from person to person. For example, it can be spread to parents changing the nappies of infected children and to workers in crèches or nurseries by similar means. Children in crèches or nurseries are also at risk of cross contamination. However, this is less of a problem where staff and children are encouraged to wash their hands thoroughly, especially after using the toilet or changing nappies.

Giardia can form spores and so can exist outside the body for long periods of time. This is what makes the spread of it so difficult to control.

Symptoms and Treatment

The *Giardia* parasite attaches to the gut lining in the upper part of the small intestine; in other words, it affects the duodenum and jejunum, as indicated in Figure 5.

Figure 5: The Digestive Tract

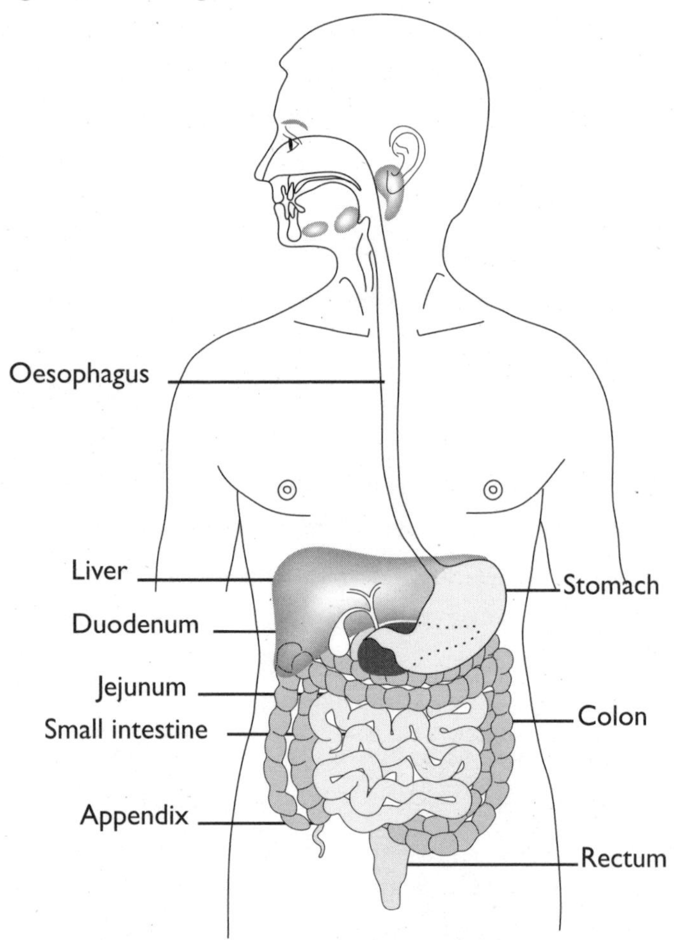

As a consequence, most of the symptoms are gut related and include nausea, pain, cramps, bloating and belching, combined with episodes of watery diarrhoea. These symptoms are often associated with a general feeling of malaise and fever.

Symptoms usually last for anything from two to four weeks and usually resolve partially or completely without treatment. However, in some patients the symptoms do not resolve and so treatment is required. The two most common drugs for treating giardiasis are both antibiotics – metronidazole (Flagyl) and tinidazole. They are both broad-spectrum antibiotics and so can damage the bacterial flora in the process of trying to eradicate the parasite. This is why the symptoms may get worse, or persist for a lot longer, in some patients.

Diagnosis

Giardiasis is diagnosed on the basis of a history of the symptoms mentioned above plus a positive result on examination of the stool. Good laboratories will usually request three stool samples as the parasite can be easily missed if only one sample is sent. The stool sample is examined under the microscope and if the spore form (dormant form) or living form is seen – see Figure 6 – then a diagnosis of giardiasis can be made.

Figure 6: Giardia

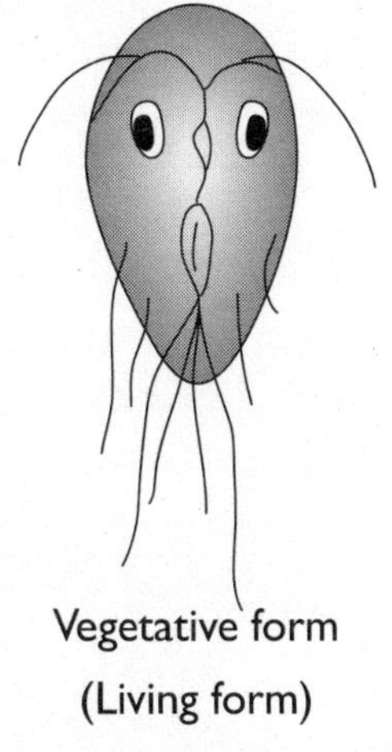

Vegetative form
(Living form)

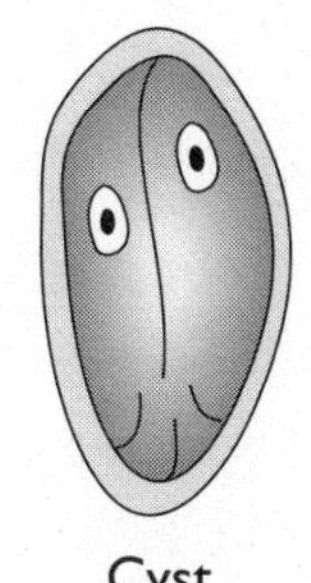

Cyst
(Dormant form)

Complications

Infection with *Giardia* can cause damage to the villi that line the gut. It can cause a flattening of the villi and so impair absorption of food across the gut wall. Usually the gut repairs itself after the infection resolves, but sometimes this does not happen and the person ends up with chronic malabsorption. This is particularly damaging in young children as it can impair normal growth and development. Chronic malabsorption can be diagnosed by means of a simple urine test called a gut permeability test. I tend to do this test routinely on patients with abdominal symptoms.

Another complication associated with this parasite is dehydration. Infection with *Giardia* can cause bouts of severe diarrhoea with very watery stools and so can result in the loss of water and electrolytes, leading to dehydration. It is especially important to look out for dehydration in cases where diarrhoea persists. It is corrected by increasing water intake and using an electrolyte solution such as Dioralyte, which is available from all pharmacies. If the dehydration is severe, the patient may require an intravenous drip.

Giardiasis also causes food intolerances. When your gut wall becomes inflamed and the inflammation lingers for weeks or months, you can start to develop intolerances to certain foods. The most common intolerance that can develop from giardiasis is an intolerance to cow's milk and the sugar in cow's milk or lactose. Lactose intolerance results in diarrhoea every time the person consumes cow's milk. However, it is possible to develop other food intolerances from *Giardia* as well.

Therefore, if you have had a bad case of diarrhoea followed

by intolerances to a number of foods, it may be worth your while getting checked for the presence of *Giardia*. I have discussed the range of gut function tests that can be carried out in my book *Hard to Stomach* (2002).

There are many parasites that can affect the gut but *Giardia* is by far the single most important in terms of the potential damage that it can do.

Ampicillin (tradename Penbritin) can favour the overgrowth of *Clostridium difficile*, which as you know can cause severe colitis, resulting in blood and mucus in the stool. Erythromycin can favour the overgrowth of streptococcal species. Because of this, it is important to wait before treating ear infections and sore throats in young children with antibiotics such as erythromycin – tradenames Erythroped, Erythrocin and Erymax. Like ampicillin above, erythromycin can also cause severe colitis.

Many antibiotics damage the intestinal flora. One in particular, known as lincomycin, can wipe out all the bacteria in the gut that use oxygen. Because it is so toxic it's rarely used today. So, although antibiotics can save your life, many of them have the potential to do enormous harm. They can predispose you to infection by parasites, which can impair your health.

ANTIBIOTICS CAN LEAD TO LOSS OF NUTRIENTS

The trillions of bacteria in the gut are important for the digestion and absorption of food, as well as the elimination of waste. However, they have some other pretty important jobs as well. They manufacture a range of vitamins, including some of the B vitamins – B1, B2, B3, B12 and folic acid – and vitamin K. Damaging the flora will result in a loss of these nutrients and may render your body deficient. The B vitamins are mainly involved in energy production, while vitamin K plays a critical role in blood clotting.

In addition to manufacturing vitamins, the gut flora also produces something else that is quite amazing. It is able to digest some of the fibre in foods such as beans, lentils and legumes, and to convert this fibre into short-chain fatty acids, which have amazing benefits for you. These acids are called, individually, acetic, propionic, butyric, lactic, hippuric and orotic acid. The most important of these is probably butyric acid as it is the main food for the cells that line the colon. Without sufficient butyric acid, the lining of the colon becomes inflamed and so colitis begins. As mentioned, a disturbance in the gut flora is now known to play a major role in the onset of colitis. As a consequence, most good gastroenterologists now prescribe probiotics for patients with this condition. Butyric acid is also known to be a key substance in the prevention of bowel cancer and possibly other cancers such as breast, prostate and liver cancer.

Some of these acids produced by the bowel flora combine to create an acidic environment that inhibits the growth of nasty bacteria such as those that cause typhoid fever. This acidic environment not only prevents the growth of nasty invaders but also helps to prevent the normal constituents of the bowel, such as *Candida albicans*, from overgrowing.

The benefits of these acids are astounding and their effects are felt far beyond the gut. They have widespread positive effects on virtually all aspects of body function. For example, propionic acid helps balance hormone levels via its effects on the liver.

The bacteria that live on you and inside you ensure you remain healthy. They act as an interface between you and your environment. To go and damage them by taking antibiotics unnecessarily is a real shame. It leads to a loss of essential nutrients, bowel problems, a predisposition to parasites, a predisposition to food allergies and the overgrowth of microbes such as *Candidia albicans* and *Clostridium difficile.*

HOW ANTIBIOTICS CAN AFFECT THE LIVER

I am now going to briefly discuss the effects that antibiotics can have on your liver, as I am including the liver as part of the digestive tract.

The liver is the most remarkable organ in the body. It has an incredible capacity to repair itself. Alcohol, certain viruses like hepatitis B and some drugs such as paracetamol all have the ability to inflict serious injury on the liver. Despite this, the liver cells can regenerate faster and better than those of any other organ; the liver is able to overcome considerable damage.

Most antibiotics have little effect on the liver. However, some have the potential to inflict serious liver damage. The newer antibiotics, especially those synthesised in the laboratory, are more at risk of doing this. I am referring to the second and third generation of antibiotics (see Chapter 1).

The antibiotics most likely to be toxic are the quinolones – more correctly referred to as fluoroquinolones. You may know them already as having names ending in 'oxacin', such as ciprofloxacin (tradename Ciproxin). The original quinolone is nalidixic acid, which is commonly used to treat urinary tract infections. The quinolones are known to cause kidney damage and liver impairment – specifically hepatitis and jaundice. Because of the many lawsuits against the drug companies that manufacture these antibacterials, doctors tend to reserve them for more serious life-threatening infections.

If a person already has some form of liver impairment, such as hepatitis or liver cirrhosis, then there is a real risk of worsening the damage through the use of any drug, not just antibiotics. All drugs are metabolised or broken down by the liver. If the liver is less able to break down the drug, the side effects associated with that drug are more likely to occur. In people with normal liver function, side effects are less common. And even if the liver does become impaired as a side effect of taking the drug, it is much more likely to recover fully.

Antibiotics have the ability to cause temporary liver dysfunction. Figure 7, showing a liver function test result, illustrates what I mean by temporary liver dysfunction, which occurs when liver enzyme levels are raised. Some of the liver enzymes referred to in the report as AST and ALT are raised above normal levels, as you can see. Once the drug was stopped, the liver enzymes returned to normal.

Augmentin is a commonly used antibiotic that causes temporary liver dysfunction. Some of the cephalosporins such as cefaclor (tradename Distaclor) can cause transient hepatitis and jaundice. Bleeding due to interference with blood clotting has been associated with all of the cephalosporin antibiotics. Antibiotics commonly prescribed for urinary tract infections can cause liver impairment. You may be familiar with these drugs by the following names: Macrodantin, Furadantin and Macrobid. These are the tradenames for the antibiotic nitrofurantoin. They are used to treat acute urinary tract infections, as well chronic infections, and can also be used as a prophylaxis to prevent the recurrence of urinary tract infections.

Figure 7: Lab Report – Liver Function Test Result

Hospital Number: ________________

Name: ________________

DOB ________________

Test Name	Result	Reference Range
ALT	40	5–33 U/L
AST	35	5–30 U/L
Total Bilirubin	5	<21 umol/L
Alk phos	63	30–130 IU/L
Gamma GT	24	3–40 U/L
Total Protein	71	60–80 g/L
Albumin	45	35–50 g/L

Nitrofurantoin is the probably the most common cause of drug-induced liver disease. It can cause a temporary elevation in liver enzymes (see Figure 7), which usually return to normal on stopping the drug. However, where the drug is used in chronic cases or used long term as a prophylaxis, more serious damage can occur. It can cause liver cirrhosis and even terminal liver failure. In other words, it can kill. It is estimated that the incidence of liver damage with this drug is approximately 1 in 1,500.

Case History: Gina, Urinary Tract Infection

Gina presented to the gynaecological outpatient department of a busy hospital with the symptoms of a urinary tract infection. She was 36 weeks pregnant and was in good health. On examination, all seemed well with both mother and baby. A urine sample was sent to the laboratory, which confirmed an infection with E.coli, a common cause of infection of the urinary tract. She was started on nitrofurantoin and discharged. A week later she returned complaining of fever, skin rash and feeling very tired and lethargic. Tests revealed raised liver enzymes and so the drug was discontinued. Three weeks later, her liver returned to normal, her symptoms disappeared and she delivered a normal healthy boy.

Most cases are like Gina's above. The enzyme levels in the patient's liver return to normal once the drug is ceased. However, in some cases the symptoms persist, sometimes leading to chronic hepatitis or chronic fatigue syndrome. A few of these unlucky patients end up with liver failure and die.

It is wise to avoid antibiotics where possible and to avoid nitrofurantoin in particular. If you have a urinary tract infection, use cranberry extract in capsule or tablet form. I do not recommend cranberry juice as there are so many types available and the majority of these are useless. It is much wiser to use the dried plant extract.

Liver damage is a known side effect of a number of antibiotics. If you are in doubt about whether to use an antibiotic, talk to your pharmacist who can inform you of all the side effects.

Chapter 6

The Link with Diabetes

When glucose is absorbed across the gut wall it enters the bloodstream and is transported to the cells of the body where it is chemically broken down to release energy. Once glucose enters the bloodstream, the blood glucose level increases and insulin is released from the pancreas. Insulin escorts glucose out of the bloodstream and into the cells. Therefore, insulin is critical in keeping blood glucose levels within normal limits.

Before looking at the link between drugs and diabetes, it is important to understand the difference between abnormal blood sugar levels and the disease diabetes. Some substances and food items can produce a temporary alteration in the blood glucose level. As long as these alterations are minor and temporary, no harm is done. However, if the blood glucose level alters a lot, one can lose consciousness and ultimately die.

So the body's blood glucose level must be kept within certain limits. If the level drops too low (a condition called hypoglycemia), one begins to feel weak or faint. This may lead to collapse and coma. The symptoms of high blood glucose are passing water more frequently, unquenchable thirst, weakness and fatigue, blurred vision, and tingling or numbness in the hands and feet. This can also lead to collapse and coma. In either case, if the blood glucose level is not adjusted back to a normal level, the person can die.

Diabetes is a condition where the body has difficulty keeping the blood glucose level within normal limits. There are two main types of diabetes: type 1 and type 2.

Most cases of type 1 diabetes occur in children and require insulin. Since insulin can only be administered by injection, this means that most cases of childhood diabetes will require daily injections of insulin. To check that the insulin is working effectively, regular blood tests need to be taken to monitor the blood glucose level. The consequences of being diagnosed with type 1 diabetes are quite severe; they essentially mean a lifetime of daily injections and blood tests. This can be very hard for a child to accept and cope with.

When I was studying medicine in the 1970s, I was taught that type 1 diabetes occurred in children, was genetic and usually required daily injections of insulin. Adult diabetes, or type 2, was not genetic and could often be treated without insulin. However, if type 1 diabetes is genetic in origin then the incidence should remain fairly constant in each country over the course of time. But the incidence of type 1 diabetes has increased exponentially since the 1950s (see Figure 8). What's more, it is continuing to increase.

Clearly the belief that it is purely genetic can no longer hold true. There have to be environmental factors responsible for such a huge increase in the incidence of this disorder. We already know many of the factors involved. These include nitrates from fertilisers, pesticides used by farmers, air pollutants, flame retardants, heavy metals and chemicals in food.

Figure 8: Incidence of Diabetes in the UK in Children under 15 Years, 1946–1988

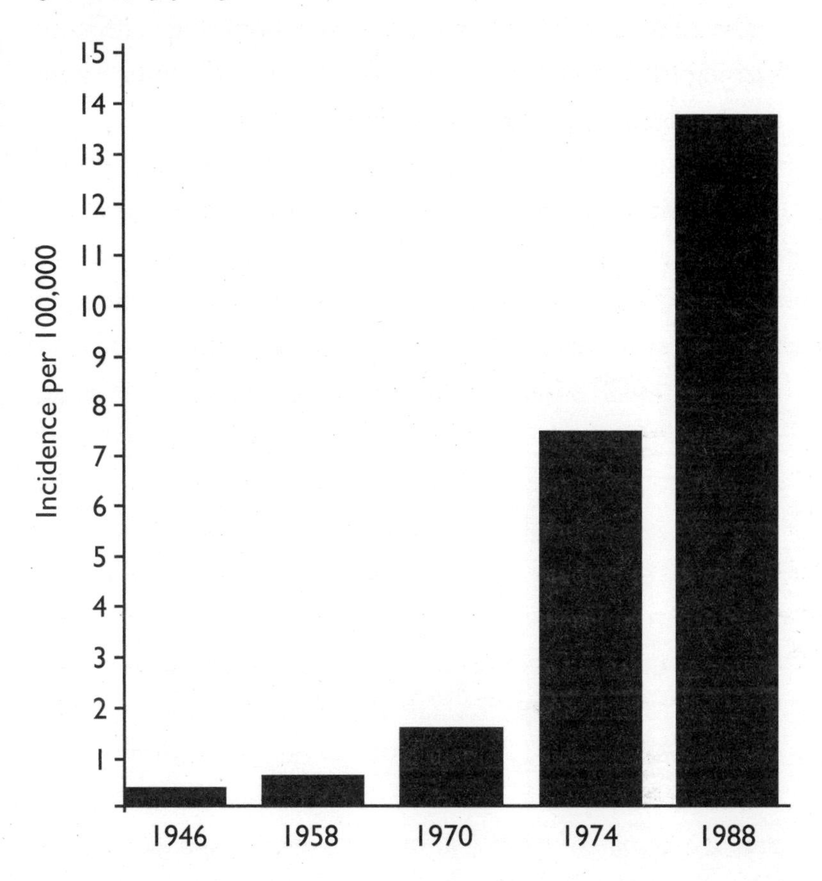

Source: Landymore-Lim, L. (1994), *Poisonous Prescriptions; Antibiotics Can Cause Asthma*, Subiaco, Australia: PODD, p. 71.

I believe conventional medicines are also partly responsible, in particular antibiotics, especially if they are used long term or are used repeatedly. Here is a case history to illustrate what I mean.

Case History: Lucy, Born 1962

Here are the GP's records for Lucy between 1964 and 1970. In this 6-year period, Lucy was prescribed a total of 33 antibiotics (identified in bold in Table 3 below). In 1970 she developed insulin dependent diabetes.

Table 3: Lucy's Drug History, 1964–1970

Prednisone	Robitussin	Phenergan
Euglate	Phenergan	Alupent
Aminophylline	Dimotane	Alupent
Achromycin	Codeine	Prednisone
Alupent	**Achromycin**	Prednisone
Prednisone	**Achromycin**	Alupent
Euglate	Prednisone	Intal spincaps
Chloromycetin	**Achromycin**	Prednisone
Achromycin	**Achromycin**	**Mysteclin**
Penbritin	Prednisone	**Achromycin**
Achromycin	Ephedrine	**Ledermycin**
Aminophylline	Prednisone	Prednisone
Achromycin	Aminophylline	Alupent
Prednisone	Alupent	Phenergan
Prednisone	Robitussin	Erythroped
Dimotane	Ephedrine	Phenergan
Prednisone	Ephidrine	Phenergan
Prednisone	Ephedrine	Aerotrol
Alupent	Robitussin	Phenergan

Ephedrine	Alupent	Phenergan
Achromycin	Phenergan	Robitussin
Dimotane	Robitussin	**Achromycin**
Prednisone	Robitussin	**Achromycin**
Prednisone	Choledyl elixir	**Achromycin**
Adrenaline	Robitussin	Robitussin
Aminophylline	**Achromycin**	Alupent
Achromycin	**Achromycin**	Robitussin
Achromycin	**Achromycin**	Robitussin
Prednisone	**Achromycin**	Robitussin
Achromycin	Alupent	**Achromycin**
Ephedrine	**Achromycin**	Alupent
Sample of **oxytertacycline**	Robitussin	Penidural
Euglate	**Achromycin**	L-codeine
Alupent	Ephedrine	Penidural
Alupent	Alupent	Avomine
Euglate	Solfex	Dimotane
Achromycin	Alupent	**Achromycin**
Achromycin	**Achromycin**	Prednisone
Alupent	**Ledermycin**	Robitussin
Achromycin	Ephedrine	Alupent
Prednisone	Alupent	Ephedrine
Prednisone		

Source: Landymore-Lim, L. (1994), *Poisonous Prescriptions; Antibiotics Can Cause Asthma*, Subiaco, Australia: PODD, pp. 97–8.

Unfortunately Lucy's was not an isolated case. Hers was one of a number of case histories of children who had been prescribed numerous courses of antibiotics and ended up being diagnosed with diabetes.

It is very distressing to read case histories such as Lucy's as they suggest that children are being damaged by the senseless overuse of drugs. There is no wisdom or intelligence at work here. These case histories cry out for a less scientific and a more caring, compassionate approach to medicine.

DRUG COMPANY ADMITS NEW ANTIBIOTIC CAUSES DIABETES

In December 2005, the drug company Bristol-Myers Squibb was forced by the Canadian government to issue a health warning to all health professionals in Canada about its new antibiotic gatifloxacin (tradename Tequin). It had to explain that the drug had the potential to disturb blood sugar levels significantly, which could result in death. In July 2003, the Health Canada website (www.hc-sc.gc.ca) published a full report alerting medical professionals to the adverse effects of the drug in its *Canadian Adverse Reaction Newsletter* (Health Canada, 2003).

Here we have clear evidence of the link between a particular antibiotic and diabetes. In this case it is not a suggestion or a theory; rather, it is a fact.

This opens up a debate about the potential for other antibiotics to cause diabetes, especially childhood diabetes. Let's see if there is any evidence linking other antibiotics to diabetes.

DO ANTIBIOTICS CAUSE DIABETES?

Gatifloxacin was not the first antibiotic shown to disturb blood glucose levels and, either temporarily or permanently, cause diabetes. An antibiotic called streptozotocin is capable of damaging the cells in the pancreas that produce insulin. This is so well known that it is one of the drugs used by scientists in animal experiments to study diabetes. So, if you are doing research on new drugs to treat diabetes and want to study the effects of these new drugs on animals, such as laboratory mice, you can induce diabetes in these mice by giving them an antibiotic. Strange but true.

Streptozotocin is not used to treat infections in humans but I mention it to illustrate the point that antibiotics have the potential to cause diabetes.

The most recently produced antibiotics are the family of drugs called fluoroquinolones or quinolones, which I mentioned in the previous chapter. A number of these drugs, such as Norfloxacin and Moxifloxacin, are known to cause alterations in blood glucose levels and diabetes.

Gatifloxacin is also a quinolone. I mentioned in Chapter 1 that there are a number of lawsuits pending against some of the drug companies that manufacture these drugs. The fact that a number of them have the potential to cause diabetes has been overshadowed by more serious side effects.

However, the important question is: do commonly used antibiotics such as penicillin and cephalosporin cause diabetes?

If you bear with me, I'd like to explain some chemistry. Insulin is made in certain cells of the pancreas. It is held in solution in these cells by binding to zinc. If it does not bind to zinc, it will not be held in solution and so will precipitate out and become useless. So, the zinc–insulin bond is very important. Anything that disturbs this bond has the potential to render insulin useless and so disturb blood glucose levels.

Certain drugs are known to break the zinc–insulin bond. How do they break the zinc–insulin bond? Take a look at Figure 9, which shows the chemical structure of streptozotocin.

Figure 9: Chemical Structure of Streptozotocin

CH_2OH

H O H

OH H

HO OH

H NH

C=O

H_3C—N— =O

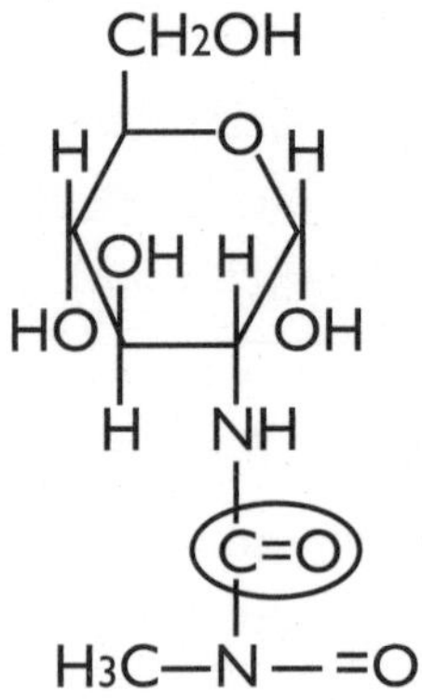

Look at the lower part of the diagram. The carbon atom (C) is connected by a double bond to the oxygen atom (O) – I have circled it in the diagram. This is called a carbonyl group and is written as C=O. Note that this carbonyl group is flanked on both sides by a nitrogen atom (N). What is of interest here is the fact that both carbonyl groups and nitrogen atoms are quite reactive as they are negatively charged.

Being negatively charged, they are attracted to positively charged atoms such as zinc and have the potential to pull zinc away from its bond to insulin. So, any drug that has a negatively charged part may be able to disrupt the zinc–insulin bond and cause diabetes.

Now look at the chemical structure of amoxycillin shown in Figure 10.

Figure 10: Chemical Structure of Amoxycillin

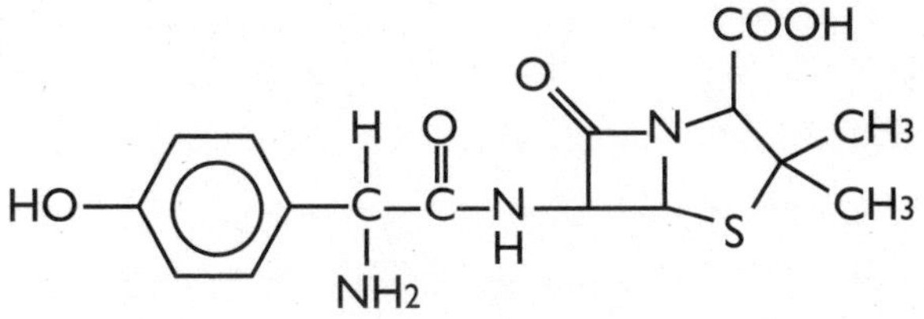

Do you see the carbonyl (C=O) group and the nitrogen atoms on both sides? This is another example of a negatively charged part of a molecule. Because the nitrogen in this case is bound to hydrogen it is less negatively charged. However, it may still have the potential to disturb the zinc–insulin bond.

Amoxycillin is a very commonly used antibiotic. The structure of most penicillins and cephalosporins is very similar to that of amoxycillin.

So, in theory at least, the commonly used antibiotics have the potential to upset blood sugar levels by disturbing the bond between zinc and insulin. If used for a long period of time or repeatedly, they may well disturb blood sugar levels sufficiently to cause diabetes.

With this in mind, let's take a look at the medical and pharmaceutical research literature for evidence that penicillins can bind to zinc. An article in the *Journal of Pharmacy and Pharmacology* in 1966 shows that all the penicillins have a tendency to bind to zinc (Niebergall *et al.*, 1966). Two other journals show that the experimental inducement of diabetes in laboratory animals with drugs such as streptozotocin can be reversed by giving the animals zinc, so confirming the theory that these drugs may cause diabetes by breaking the zinc–insulin bond (Lazarow *et al.*, 1951) (Tadros *et al.*, 1982).

This is very concerning information. It suggests that a number of commonly used antibiotics such as ampicillin (Penbritin) and amoxycillin (Amoxil), which are broad-spectrum penicillins, have the potential to cause diabetes. Not to say that everyone that uses these drugs will develop diabetes. However, if such a drug is used repeatedly or continuously for a period of time, there may be a risk of developing disturbances in blood glucose levels. Is this why diabetes is becoming more common in young children? It could well be one of the factors responsible.

However, is there any evidence to suggest a link between the frequent use of antibiotics and the incidence of diabetes? Unfortunately, such evidence exists and it comes from an unusual source – a chemist.

Dispensing Medical Practices in the UK

Dr Landymore-Lim studied Chemistry at the University of Sussex, Brighton, UK. She became interested in how certain drugs alter body chemistry. Studying childhood diabetes, she began to look at possible risk factors by examining the drugs mothers were exposed to during pregnancy and delivery, and the drugs the babies were exposed to prior to the onset of diabetes.

From her research it became clear that there was an association between the child being exposed to antibiotics and the development of diabetes. What I found particularly interesting was the clear association between a certain type of medical practice and the incidence of diabetes. Let me explain.

In rural areas of the UK, doctors are allowed to dispense drugs if there is no pharmacy within easy reach of the patient. In other words, doctors are paid extra to play the role of pharmacist. They have the potential to supplement their income in this way. It is therefore a system that is open to abuse.

Some doctors may be tempted to over-prescribe, thereby exposing children to the potential side effects of drugs. By far the most commonly prescribed drugs for young children are antibiotics. If antibiotics are a factor in the development of childhood diabetes and if doctors are over-prescribing these drugs in rural areas of the UK, one would expect a higher than normal incidence of childhood diabetes in these areas.

Dr Landymore-Lim carried out research in the Cambridgeshire and Wessex Area health authorities. She found that rural areas where there were dispensing medical practices had a much higher than expected percentage of children with diabetes. In Figure 11 you can see that the place with the highest incidence is East Anglia, a very rural part of England.

It is extremely odd to find isolated areas of the UK with such a high incidence of diabetes. Further evidence of this result is found in other isolated areas of the UK. An article in the journal *Diabetologia* in 2006 indicates that there is a higher than expected incidence of type 1 diabetes in remote areas of Northern Ireland (Cardwell *et al.*, 2006).

Figure 11: Distribution of Regions of Varying Incidence of Diabetes in Under-15s in England, Occurring during 1988

Source: Metcalfe, M.A. and Baum, J.D. (1991), 'Incidence of insulin dependent diabetes in children aged under 15 in the British Isles during 1988', *British Medical Journal*, vol. 302, no. 6774, p 111.

So, more children are developing type 1 diabetes in rural areas of the UK compared to urban areas. Clearly something unusual is happening. Maybe this pattern is reflected in other countries as well. However, this is not the case. This may be because most countries do not have dispensing medical practices, as in the UK, or they may be more cautious in their use of antibiotics in young children.

If a disease is caused by the use of conventional drugs then one would anticipate a higher than expected incidence of the disease in areas where there is higher usage of these drugs. It follows that one can anticipate a higher than expected incidence of the disease in areas where there are dispensing practices and where doctors are overusing such drugs.

If antibiotics are causing diabetes then countries with a high usage of antibiotics will have a high incidence of diabetes, and those countries with a low usage will have a low incidence of the disease.

This is exactly what we find. Figure 11 showed a map with the incidence of type 1 diabetes in different regions of the UK.

Figure 12 is a map of the regions of the UK with a high density of dispensing medical practices.

The two maps are strangely similar. For example, East Anglia has the highest incidence of type 1 diabetes and also the highest number of dispensing medical practices. This is a very worrying pattern. At a minimum it requires investigation by the health authorities.

By comparison, there are no dispensing medical practices in the Republic of Ireland and the incidence of type 1 diabetes is low. This is further evidence in favour of the link between antibiotic use and type 1 diabetes.

The suggestion that the liberal use of antibiotics may be one of the causes of type 1 diabetes may help to explain not only why different areas of the same country have different diabetes incidence rates but also why different countries have such different incidence rates.

Figure 12: Distribution of Regions of Varying Density of Dispensing Medical Practices in England as of 1 October 1989

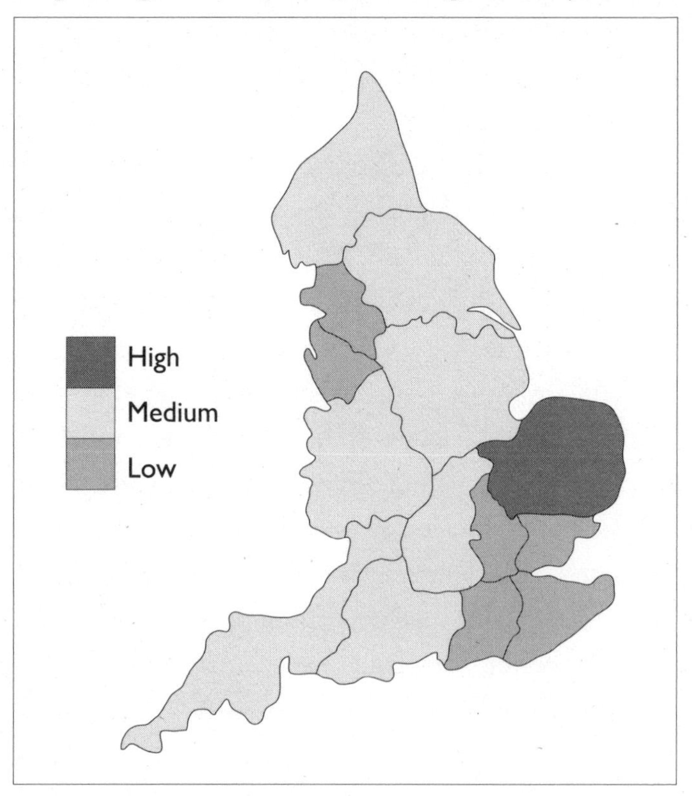

Source: Landymore-Lim, L. (1994), *Poisonous Prescriptions; Antibiotics Can Cause Asthma*, Subiaco, Australia: PODD, p. 111.

When one looks at a map of the world showing the incidence of type 1 diabetes, it is interesting to find the northern European countries such as Finland, Sweden, Norway and the UK top the list. There should be further studies done to try to explain why the incidence of diabetes varies so radically from 57 per 100,000 of the population in Finland to 0.1 per 100,000 of the population in Venezuela.

Clearly the reasons are varied and involve not only genetic factors but a number of environmental factors, including possibly the use of drugs such as antibiotics.

Some antibiotics are known to cause diabetes. Others, such as the commonly used penicillins and cephalosporins, may well have the potential to do so. One particular antibiotic is used in laboratory animals to induce diabetes.

As Dr Landymore-Lim's research in the UK suggested, exposure to antibiotics in the womb or in early childhood may be a risk factor for the onset of type 1 diabetes. This early exposure to these drugs may disturb the chemical bond between zinc and insulin, breaking the bond and rendering insulin useless.

Dr Landymore-Lim's research has uncovered something else which may well require further investigation. She uncovered a link between the incidence of type 1 diabetes in young children and dispensing medical practices in rural parts of the UK.

There is enough clinical and experimental evidence to carry out much further investigation of these links. However, such research would have to be carried out by those with no links to drug companies and there would have to be no conflicts of interest. It would need to be funded by public money.

Chapter 7

Antibiotics and Vaginal Thrush

Most women experience vaginal thrush at some stage in their lives.

Antibiotics can upset both the gut and vaginal flora. The use of antibiotics can result in chronic infection, especially if the antibiotic is used repeatedly or over a long period of time, as in the treatment of acne. Many cases of acne that I have seen over the years can be complicated by intestinal dysbiosis and vaginal thrush.

It is commonly known that thrush presents as vaginal irritation, itching and in some cases a whitish discharge. Most cases are quite simple to treat; others, however, can be very persistent, requiring a much broader approach. Let me explain.

Case History: Maria, Vaginal Itching

Maria had symptoms of vaginal itching and some discharge that stained her underclothes. Her general health was very good and she had not been on antibiotics and was not using the contraceptive pill.

She did not wish to use a conventional antifungal drug, as she preferred to use a more holistic approach. I suggested she use a probiotic supplement orally and locally. As regards the latter, I asked her to break open the probiotic capsule and use the powder to coat the wall of the vagina by means of a tampon. Within a few days her symptoms were gone, and they did not come back.

I am using Maria's case history to illustrate a few points. First, most cases of vaginal thrush are limited to the vagina and so local treatment will work. Second, vaginal thrush is very easy to treat; it does not require drugs but rather some friendly probiotics. It is possible to use the probiotic locally by coating a tampon with the probiotic powder or applying the probiotic powder with a finger to the vaginal wall.

It is best that the person avoids sexual intercourse for the duration of treatment in case of cross contamination, as the next case illustrates.

Case History: Angela, Recurrent Thrush

Angela was having recurrent bouts of vaginal thrush. Each time that she treated it the symptoms abated, only to resurface again days later. This usually suggests that the problem is not only vaginal but may be systemic. She did not have any other symptoms and was not on any medication. She then told me that her husband was complaining of inflammation of the tip of his penis. I then suggested she bring her husband to see me. Sure enough, he had the signs of penile thrush.

I treated both of them simultaneously with probiotics applied locally and told them to avoid intercourse.

Within a week, both were symptom free and they remained so.

In all cases of vaginal thrush it is important to remember to consider the male partner, if there is one. It is necessary to treat both partners at the same time to prevent them from re-infecting each other. Otherwise the problem will become recurrent, as was the case with Angela.

The wall of the vagina is coated with millions of bacteria, including yeasts. These normal constituents of the vagina keep the vaginal secretions acidic, so preventing infection. Anything that alters the bacterial population, such as antibiotics, or anything that alters the acidity of the vaginal secretions, such as menstruation, can result in thrush.

Pregnancy alters the ecology of the vagina, mainly because of hormonal changes. Therefore, vaginal thrush is common during pregnancy. In this case, it is best to treat the thrush locally and naturally to avoid the baby being exposed to conventional drugs. However, as the next case history illustrates, it can be difficult to treat.

Case History: Jill, Recurrent Thrush

Jill was in the first trimester of pregnancy when she came to see me complaining of the classical symptoms of thrush. Her gynaecologist had sent a swab to the laboratory and confirmed the presence of Candida albicans, *a yeast. He prescribed a conventional antifungal treatment, but she preferred to be treated naturally and so opted to come to me.*

Initially she responded well to treatment, but within a few weeks the thrush came back. Knowing that pregnancy can alter the acidity of the vagina, so allowing yeast to overgrow, I decided to repeat the treatment. Within two weeks, she was back again. I was now suspicious that there was something else going on.

A urine test revealed the presence of glucose, which can also happen during pregnancy. It seemed the glucose was disturbing the vaginal ecology. I suggested that Jill persist with treatment and have her blood and urine monitored throughout the pregnancy.

After the birth of the baby, her urine test returned to normal, as did the flora of the vagina. She did not suffer from thrush again.

The natural treatment of vaginal thrush has been shown to work in scientific studies. There is a wealth of research showing the benefit of probiotics in treating not just thrush but other vaginal infections. The research suggests that both oral supplementation (swallowing a probiotic capsule) and local application (see above) work well. To make treatment more effective, I usually recommend doing both.

Basically, good bacteria in the probiotic powder counteract or replace bad bacteria and yeast that tend to overgrow if the flora is disturbed. Slowly the good bacteria reduce the number of overgrown bacteria to a sufficiently low level where balance is restored and symptoms disappear.

Here is a quick look at some of the research carried out to show the effectiveness of probiotics in the treatment of vaginal infections such as thrush.

Sixty women who had the symptoms of a vaginal infection were treated with either a vaginal suppository containing *Lactobacillus acidophilus* (*L. acidophilus*) or a suppository containing *L. acidophilus* and *L. paracasei*. After three months of treatment, both groups showed a significant improvement in symptoms as well as a significant improvement in vaginal odour and acidity of the vaginal secretions (Delia *et al.*, 2006).

In a study published in 2007, researchers in the University of Milan treated 40 women who had a vaginal infection with a vaginal douche containing *L. acidophilus* for a short period of 6 days. After treatment, only 3 of the women still had the same symptoms, all had lost the typical odour of a vaginal infection and 34 out of the 40 women had a normal acidity of the vaginal secretions (Drago *et al.*, 2007).

There are many studies from different countries such as Israel, Sweden, Brazil, Canada and Austria, all illustrating the beneficial effects of probiotic treatment. Most of these studies look at short-term treatment, such as six or seven days, but a few use longer periods such as three months. Because the body has trillions of bacteria, it may take time to alter the population, so it is best to use a probiotic supplement for at least three months, especially if you have used antibiotics in the past. In truth, one should really take probiotics every day to replenish the good bacteria that die off naturally.

Most cases of vaginal infection are no more than local infections and require no more than probiotic treatment. However, occasionally these infections are complicated by a systemic problem. Where this is the case, treatment should address the systemic problem as the local infection will not improve until this is sorted out.

WHAT ARE THE OTHER CAUSES OF VAGINAL INFECTION?

Let's look at some other factors that can result in a vaginal infection, as part of the bigger picture.

Drugs

Apart from antibiotics, conventional medicines such as the contraceptive pill can result in a chronic vaginal infection. The contraceptive pill alters the hormonal system that controls the vaginal secretions. The oestrogen component of the pill also alters the gut flora. So 'the pill' can cause intestinal dysbiosis along with vaginal infection.

Diabetes

As you have learned already, diabetes can result in high levels of glucose in the bloodstream and in urine. Because of the female anatomy, glucose in urine can alter the vaginal flora and favour the growth of yeasts such as *Candida albicans.* So, a systemic problem such as diabetes can result in vaginal thrush.

Weakened Immunity

Anything that impairs your immunity such as chronic stress can impair your first line of defence, the bacterial flora, so resulting in intestinal and vaginal dysbiosis. As mentioned in Chapter 4, the main factors that impair immunity are:

- Poor diet
- Ongoing stress
- Drugs such as steroids and antibiotics, drugs used to treat auto-immune disorders, and immunosuppressive drugs such as azathioprine

Metals

Metals cause many disturbances, principally in the gut and the nervous system. However, some metals can disturb the hormonal system as well, resulting in chronic thrush, which is very resistant to treatment. The following case history illustrates this point.

Case History: Siobhan, Chronic Thrush

Siobhan had been suffering from thrush for almost two years. None of treatments she had tried worked, including many conventional drugs and many forms of natural medicine. She came to me as she was desperate to find the underlying cause. All the tests done previously did not indicate an underlying problem.

On questioning her, I learned that she was told by her dentist that she was grinding her teeth during her sleep, something she had not been aware of. She also had a number of large amalgam fillings. I became concerned that she was grinding the metal in her fillings and the resulting dust was being swallowed into her gut. Amalgam fillings also release mercury vapour, which can be inhaled (amalgam is 50 per cent mercury).

We did tests on Siobhan and discovered she was indeed excreting toxic levels of mercury in her stool and urine. I then referred her to an amalgam-free dentist to have her fillings replaced. After all the metal was removed, I then treated her using oral chelation and high levels of antioxidants.

Repeat tests done many months later revealed a significant reduction in the level of mercury that Siobhan was excreting. I continued Siobhan's treatment for a further six months, during which she used probiotics both orally and vaginally.

Siobhan's vaginal thrush symptoms abated and the acidity of her vaginal secretions returned to normal for the first time in three years. Since then, she has not suffered from thrush.

Siobhan's was a very unusual case: most cases of mercury toxicity do not present with chronic thrush. They usually present with low energy, gut problems or neurological problems. It was unusual to find that underlying the vaginal flora problem was a systemic flora problem, and underlying that was a toxicity problem.

I ask all patients about clenching or grinding teeth and also check to see the size of their fillings and the number of fillings and to see if they have been worn away. I am shocked that dentists in Ireland and the UK continue to use amalgam fillings, despite the fact that mercury is a highly toxic substance and releases a vapour continuously. Dentists love mercury because it is liquid at room temperature and so will fill a cavity with ease and will bind well to the tooth. It is also dirt cheap in comparison to composite fillings (white fillings), which don't contain metals.

If mercury is inhaled or swallowed into the gut, it disturbs the bacterial flora of the body. Since some of it may be excreted in urine, it can ultimately cause a disturbance in the vaginal flora as well. This was what happened in Siobhan's case.

In summary then, vaginal thrush is usually uncomplicated and easy to correct with a good probiotic. Occasionally, thrush is part of a bigger health problem and requires a more detailed history, deeper examination and further tests to find the root cause.

Chapter 8
Antibiotics and the Immune System

Everyone is aware that when your immunity is low you are more prone to infections and therefore may need an antibiotic. What is less well known is the effect that antibiotics can have on different parts of the immune system. Before exploring this, it is necessary to explain the different parts of the immune system and how they interact.

PARTS OF THE IMMUNE SYSTEM

The Bacterial Flora

Starting from the outer layers of the body and working our way inwards, the first part of the immune system we encounter is the bacterial flora that coats all the surfaces of the body, including the skin, airway, gut and vagina. Because they are tiny microscopic organisms, we do not see them. However, there are 10 bacterial cells for every human cell; it's no exaggeration to say that they are important. In actual fact, they form the single most important part of your defences against bugs and toxins.

If you asked people to name parts of immune system, they would most likely mention the white blood cells, antibodies or killer T-cells. I suspect very few people would mention the bacterial flora, as we have been educated to link the word 'bacteria' with the word 'infection' and not with the word 'immunity'. We use antibacterial wipes, detergents, toilet cleaners, etc., so reinforcing the link between bacteria and infection. The truth is rather different: most bacteria are friendly fellows and are critical for life on this planet to survive.

Trillions of bacteria coat the surfaces of all living creatures to protect them from their environment. To maintain a high population of bacteria in your body, it is important to take a probiotic supplement, live yoghurt or soured milk (buttermilk) everyday so that dead bacterial cells are replaced by living ones. Good bacteria keep this population in check so that opportunistic bacteria such as *Staphylococcus spp.* do not overgrow. Good bacteria also counteract harmful bacteria and prevent them from taking hold and setting up an infection.

This bacteria population or flora acts as a dynamic, ever-changing interface between you and your environment. It is there to protect you and keep pathogens at bay. It forms a critical part of the immune system.

The Skin

The next barrier that protects the body from harm is the skin. As we know, this coats the outside of the body, forming a defence against harmful bacteria, chemicals, etc. in the outside world. It also acts as a cushion against physical injury, especially if there is a lot of subcutaneous fat.

The Mucus Membranes

The next layer that acts as a defence consists of the mucus membranes that line the airway, the digestive system and the vagina. As the name suggests, these membranes produce mucus, which entraps harmful bacteria, viruses or particles, and expel the mucus by coughing and sneezing. As mentioned, bacteria line these surfaces as the primary defence layer; the act of producing mucus is the second line of defence. This is usually sufficient to deal with whatever the environment throws at you. If these barriers do not function effectively then the surface layer may become irritated and inflamed.

Inflammation of the skin or mucus membranes may be a sign that the external barriers have been breached. Smoking cigarettes can irritate the airway and cause inflammation, resulting in bronchitis. Skin inflammation, skin rashes and skin itching are all associated with auto-immune disorders such as coeliac disease – a type of dermatitis called dermatitis herpetiformis is part of coeliac disease – with bacterial or viral infections such as measles, and with fungal overgrowths such as candidiasis.

So the outer layers of the body are protected by bacterial flora, the skin and the mucus membranes. Now let's take a look at the internal immune system.

The White Blood Cells

There are a number of different kinds of white blood cell, each with a slightly different function. First, there are *macrophages*, which patrol around the nooks and crannies of the body hunting for foreign invaders such as bacteria or viruses that may have gotten through the outer layers of defence. Macrophages engulf any abnormalities, so destroying them.

Next there are the *neutrophils*, which are white blood cells that patrol within the bloodstream checking for abnormalities among various cells. Once they identify a problem, they signal for help to other parts of the immune system and begin to clear debris from the area.

The most important type of white blood cell is probably the *lymphocyte.* It is also the most sophisticated in that it has the ability to retain a memory bank of harmful substances so that it can mobilise resources more rapidly on exposure to such substances again. There are two basic types of lymphocyte: the *B-lymphocytes* (B-cells), which are produced in the bone marrow and modified in the spleen, and the *T-lymphocytes* (T-cells), so called because they are modified by the thymus gland. Both types of lymphocyte circulate via the lymph channels, blood stream and tissue fluid.

B-cells produce antibodies. These are proteins that help entrap dangerous microbes or toxins before they get the chance to invade the cells of the body. These antibodies are designed to attach to a specific virus, bacterium or toxic chemical and disable it, allowing other parts of the immune system, such as the neutrophils, to finish the job of destroying it.

T-cells are more focused on protecting the body's cells after invasion. For example, when a harmful virus invades a cell in your body, T-lymphocytes mobilise and may destroy the whole cell in an attempt to destroy the invading virus. These T-cells that kill cells as a form of protection are aptly named *killer T-cells.* They are able to kill selected cells by inserting special chemicals into the cells or they instruct the cells to self-destruct.

There are also *helper T-cells*, which are involved in scanning the cells of the body for damage by an invading microbe or toxin. When these helper T-cells detect a problem, they signal to the killer T-cells to come to the defence. Helper T-cells also carry a memory of previous invasions and help to coordinate battle plans.

T-cells and all parts of the immune system have a very clever communication system similar to how we humans communicate. We use mobile phones, e-mail, Facebook and Twitter to send messages to one another. In this way we are able to plan and co-ordinate as a group. All parts of the immune system use chemicals called cytokines. These cytokines allow all parts of the body to work in a co-ordinated fashion to achieve a particular objective such as ridding the body of a dangerous virus.

For example, it is known that the bacterial flora utilise this cytokine-mediated communication to pass important information to the body's white blood cells. There is now a lot of scientific research validating this interconnectedness of all parts of the immune system (Hlivak *et al.*, 2005).

HOW ANTIBIOTICS AFFECT THE IMMUNE SYSTEM

First, antibiotics are known to damage the bacterial flora. The more broad spectrum the antibiotic, the more the flora is disturbed. Narrow spectrum antibiotics such as benzyl-penicillin cause less harm to the flora. Narrow spectrum means that only a limited range of bacteria are killed by the antibiotic, whereas broad spectrum means that many types of bacteria are killed.

Since many of the antibiotics used to treat infections today are broad spectrum, such as amoxycillin, tetracycline and metronidazole, the first line of defence of the body's immune system is often damaged during the course of treatment. This is why doctors used to advise patients to take live yoghurt while on a course of antibiotics. Today, it is best to use a well-researched probiotic supplement such as 'For Those on Antibiotics' by OptiBac. This will minimise the damage done.

Antibiotics can affect the immune system in other ways as well. Most importantly, they can impair white blood cell function. This crucial line of defence is damaged by a whole host of antibiotics.

Tetracycline antibiotics such as Minocin, Vibramycin, Terramycin, Sumycin, Achromycin and Declomycin affect all parts of the immune system. They slow down the movement of white cells to the site of an infection, so delaying the whole process of fighting the infection and clearing up debris (Thong and Ferrante, 1980). In other words, the normal inflammatory response, which allows more blood into the site of damage or infection, is impaired.

Tetracycline antibiotics also interfere with the production of mature ready-for-action T-cells by the thymus gland. That's a bit like interfering with a soldier's training, making him/her less able to cope in battle. Like all white blood cells, T-cells are manufactured in the bone marrow. They then migrate to the thymus gland behind the breastbone, where they undergo maturation and specialisation to form helper cells, killer cells and other types of lymphocyte with specialised functions. All the tetracycline antibiotics interfere with this very important function of your immune system (Kloppenburg *et al.*, 1995).

Tetracycline antibiotics interfere with antibody production by B-lymphocytes, again making it very difficult for other parts of your immune system to identify the targets to attack.

They also interfere with the ability of white blood cells to clear up the site of an infection by removing dead cells and other debris.

Table 4: The Antibiotics that Suppress Different Aspects of the Immune System

Chemataxis[1]	**T-Cell Transformation**	**Delayed Hypersensitivity**
Tetracycline	Tetracycline	Tetracycline
Doxycycline	Doxycycline	Doxycycline
Gentamycin	Minocycline	Metronidazole
Tobramycin	Trimethoprim-Sulfa.	Rifampim
Rifampim	Clindamycin	

Antibody Production	**Phagocytosis**[2]	**Metabolism of White Blood Cells**
Doxycycline	Tetracycline	Trimethoprim-Sulfa.
Chloramphenicol	Doxycycline	Trimethoprim
Trimethoprim-Sulfa.	Amphotericin B.	Chloramphenicol
Rifampin		

[1] Movement of white blood cells to site of infection.

[2] The way white blood cells engulf and destroy microbes.

Source: Hauser, W.E. and Remington, J.S. (1982), 'Effects of antibiotics on the immune response', *American Journal of Medicine*, vol. 72, no. 5, p. 711.

Tetracycline antibiotics actually do much more than interfere with white blood cells. These antibiotics affect all the major parts of the immune system. There is much research to substantiate these claims and Dr William Hauser and Dr Jack Remington of Stanford School of Medicine, one of the best medical schools in the US, have called for the inclusion of immune system suppression as a side effect of these drugs (Hauser and Remington, 1982).

However, tetracycline antibiotics are not the only antibiotics to interfere with different parts of the immune system. Table 4 shows what antibiotics affect what parts of the immune system.

As you can see from Table 4, Rifampin, which is the trade-name for the antibiotic rifampicin, also interferes with more than one part of the immune system. Rifampin is used to treat tuberculosis (TB). People with TB tend to get the disease because their immunity is impaired; the use of Rifampin may indeed add to this impairment. You can also see from the table that the drug trimethoprim-sulfamethoxazole, sold under the trade-names Septrin or Bactrim, also has the potential to cause impairment of different parts of the immune system. The use of any of these antibiotics in patients whose immunity is already impaired demands great caution. Fortunately, most of the penicillin-type antibiotics do not cause immuneosuppression.

Let me just provide a word of caution regarding the use of drugs in the treatment of asthmatics. As you know, some antibiotics can suppress the immune system. However, steroids can also suppress immunity, especially if used long term. The combination of antibiotics and steroids is often the first line of treatment when a patient is admitted to hospital with asthma.

There are many causes of asthma, including allergies, infection, emotional distress, nutritional deficiencies and gluten sensitivity – to name a few. Regardless of the underlying cause, the same drugs are used for treatment of most cases of asthma. The case history below illustrates the danger of treating patients with the same diagnosis in the same way.

Case History: Andre, Asthma

Andre was a young man of 22 when I first met him. He had been admitted to hospital with severe asthma 10 times in the course of 6 months. Twice he died in casualty and had to be resuscitated.

When Andre came to see me, I was alarmed by two things: first, by the number of drugs he was on and how high the dosages were; second, by his number of hospital admissions. This young man was in difficulty but not just at a physical level. Andre was not aware of any emotional stresses in his life when I questioned him. The real story came from his adoptive mother at a later date.

Prior to being adopted, Andre had suffered great physical abuse at the hands of his biological parents. As a young child, he and his brothers and sisters were locked in cupboards for days at a time. They were starved, beaten and burned with cigarette butts. The children were eventually rescued by social workers and placed in care.

Since Andre was very young at the time, he was unaware of much of the abuse at a conscious level; the information was locked away in his unconscious. It did, however, trigger a pattern of behaviour when he got into difficulty in a relationship.

On further questioning, it transpired that all of Andre's

asthmatic attacks were preceded by a disagreement with a loved one – girlfriend, or adoptive mother or father. Any threat to these relationships was a threat to his survival. The emotional turmoil was probably too painful for him to deal with and may have triggered a cry for help in the form of an asthmatic attack. The words his adoptive mother used to describe his asthmatic attacks gave me the impression of a 'death wish', as if Andre found the emotional pain so difficult to deal with that he wanted to die.

Conventional medicines may have been helpful to Andre, but it was apparent that his real need was for emotional help. I am pleased to say that he is now getting emotional help, he has not seen the inside of a hospital for the past 18 months, and he has been able to reduce some of his medication and stop using the rest of it. No amount of conventional drugs or natural medicines would have solved Andre's problem.

Andre's case shows that the interaction between the emotional and physical within us must never be forgotten or underestimated. It also demonstrates the inherent danger of viewing a patient purely in physical terms and as having purely physical symptoms, which was what happened each time Andre was admitted to hospital.

Some antibiotics can suppress immunity, as can steroids, but emotional stress is also a suppressor, especially if prolonged. The use of certain antibiotics such as tetracyclines should be avoided if at all possible in patients showing signs of stress.

If you have to use an antibiotic that may suppress your immune system, or you need to use steroids for longer than two weeks, then use measures to boost your immune system.

One simple measure is to take a high dose of vitamin C, such as 1000 mg, three times daily.

WAYS TO HELP YOUR IMMUNE SYSTEM

Helping your immune system becomes very important if you work in a hospital, care home or any other place where you may be exposed to infected individuals. It's also important if you feel your defences are impaired because of the use of conventional medicines, such as steroids, certain antibiotics or immunosuppressive drugs, or because you have been stressed for a sustained period of time. However, all of us should help our immune systems through the natural means I outline here.

Again, let's start with the outer layers of the immune system and work our way inwards, looking at ways to protect ourselves from getting an infection in the first place.

Protecting the Flora

Protecting the bacterial flora is the most significant thing you can do to boost your natural defences. I am often asked by patients and by the general public during public talks to name the single most important nutritional supplement. My answer is always the same: curds and whey. The curds contain live bacteria and the whey has immune-boosting substances. Good examples of foods that contain curds and whey are live yoghurt, soured milk or buttermilk.

Every culture on the planet has used fermented milk products such as live yoghurt. I grew up drinking buttermilk and when I visited West Africa I saw that the Fulani, who herd cattle, start their day with curded milk.

Curded milk products such as live yoghurt contain the two most important species of bacteria, *Lactobacillus acidophilus* and *Bifidobacterium bifidum*, which are abbreviated to acidophilus and bifidum and can be abbreviated further to 'A' and 'B'. These two bacteria hold the secret to the therapeutic benefits of live yoghurt. You can also get these bacteria in a good probiotic supplement (see Chapter 9).

Do not confuse live yoghurt with the huge array of commercial yoghurts available on supermarket shelves, even if they add misleading words on the label such as 'made with live cultures'. Live yoghurt differs from commercial yoghurt in that it is not heat treated after the bacterial cultures are added. Therefore the bacterial cultures in live yoghurt are able to survive in the gut and multiply. Most commercial yoghurts are heat treated, which kills the beneficial live cultures. They are therefore effectively useless. If the yoghurt is left to stand for a few days and it curds then it has live cultures; if it doesn't curd, it is not a live yoghurt. This is the simplest way of distinguishing the two.

The main benefits of consuming live yoghurt or curded milk products is that they contains bacteria that produce acids, such as lactic acid, which inhibit the growth of bad microbes. Beneficial bacteria in curded milk also play a critical role in digestion and protect against bowel cancer. But probably the most important function of the body's flora is to interact with the white blood cells.

Researchers in Japan found that supplementation with beneficial bacteria increased the lymphocyte count and in particular increased the number of killer T-cells (Sugawara *et al.*, 2006). Theirs is one of many research articles showing how closely the external layers of the immune system communicate with the internal layers. In other words, as mentioned earlier in this chapter, the bacteria colonising your body are continuously sending information to your white blood cells.

Protecting the Skin

Your skin can be protected in the most basic way: by washing your hands correctly after handling animals, meat, and public surfaces such as door handles, supermarket trolleys or baskets, etc. Washing your hands has become very important in preventing the spread of highly resistant bacteria.

However, washing doesn't mean washing with water alone. Many bacteria are only killed with the action of soap and warm water. It is best to encourage good hygiene at home and at work, especially if you work close to infected animals or humans, or you work with meat products.

If your skin is breached because of a cut or wound, it is important to sterilise the cut or wound with an antiseptic and then cover it with a bandage or plaster. This is especially important for anyone dealing with blood or blood products. In my work with HIV patients, because of the risk of infection, I was taught to cover even a minor skin abrasion such as an insect bite with a bandage.

Boosting the White Blood Cells

What we eat and the supplements we take can boost the number and activity of white blood cells.

I shall now discuss the role of diet, nutritional supplements and herbs in protecting and boosting the white blood cells and the immune system as a whole.

Diet

My book *Good Food* (Gill & Macmillan, 2013) deals with diet in detail. However, here I will summarise the most salient points.

First, eat energy rich, natural foods. If you see it growing in the fields or standing in the fields, eat it. Avoid or minimise energy-poor foods, that is, the man-made foods that come in tins, bottles, jars and packets, etc. The reason for this is that your body is natural and requires natural energy to remain healthy.

All energy on this planet comes from the sun. The sun provides us with heat and light energy. Plants use the light energy to make food in a magical process called photosynthesis. This energy is then passed on to us when we eat the plants or when we eat the products of animals that feed on the plants. Put simply, the sun's energy should end up in your body, keeping it healthy.

Many man-made foods, also called processed foods, contain large amounts of sugar to guarantee that they taste good. Sugar weakens your immune system. It does this principally by disturbing the bacterial population of the body, so interfering with the first line of defence. Sugar disturbs the flora by favouring the growth of yeasts and fungi.

Sugar can also affect other parts of the immune system. Research has shown that sugar can impair the ability of white blood cells to gobble up and kill invading bacteria (Sanchez *et al.*, 1973). Other studies have shown that sugar can rob the body of certain nutrients such as zinc (Werbach, 1996), which is critical for immune function.

Therefore, sugar has the ability to interfere with two lines of defence of the immune system. It is not only devoid of energy but it can rob your body of energy. Clearly, if you wish to protect your body from infection, you should minimise the amount of sugar in your diet.

There are a lot of hidden sugars in sports drinks, energy drinks and soft drinks. It is best to avoid these and drink lots of water instead. Water is the single most important nutrient as 50–60 per cent of your body is composed of it. All of the major functions of your body require water. Yet many people do not consume sufficient quantities of it.

When you drink water, make sure that it is not tap water, which may contain chlorine, fluoride and, in some cases, heavy metals (McKenna, 2013). The quality of our water supply has become a major public health issue. Children seem to know instinctively that tap water is unsuitable for drinking – the odour and taste may give them clues. Many children in modern cities and towns take in the bulk of their fluid requirements in the form of soft drinks.

When I was a child in Northern Ireland, we lived in the countryside and had a well in our back garden. This was our main source of water for a number of years. Since few chemicals, pesticides and insecticides were used on the soil in those days, the water was pure and safe to drink. It came straight from Nature. Now, however, reports of the chemical analyses done on the water table in different parts of Europe leave one in no doubt about the damage done to our precious water supply over the past 60 years. The most basic of all nutrients, water, has been rendered unsafe. As a result, we are forced to use filtered or bottled water.

As I mentioned, food needs to be energy rich and natural. The same applies to water. The energy component of water is as important as its other constituents. This is why glacial water and artesian water from deep within the earth are the safest forms of water. The best water on the planet is glacial water from melting glaciers in Pakistan, Northern Europe, Tibet, Chile and Argentina, as well as water from artesian wells such as Spa mineral water from Belgium, which is available throughout Europe. Spa mineral water has won many awards over the years. Water direct from Nature is far superior to any other form of water as it is energy rich.

Nutritional Supplements

In this section I discuss the most important vitamin, vitamin C, the most important mineral, zinc, and the herb *Echinacea purpurea* – all of which have been shown to have a positive effect on immunity.

Vitamin C

The prime supplement to help boost your immunity is vitamin C. Most people think that citrus fruits are the best source of this nutrient, but, as Table 5 shows, many vegetables also have high levels of vitamin C. In fact, some vegetables have higher levels than many citrus fruits.

The most important fact to remember about vitamin C is that exposure to air destroys it, so eating fresh vegetables and fruits is essential. Human beings are one of the few animals unable to manufacture vitamin C in our bodies and so we depend entirely on fresh fruits and vegetables. It is therefore easy to become deficient in this vitamin. The classic symptoms of a deficiency in this vitamin are bleeding gums, poor wound healing, bruising, susceptibility to infection and depression.

Table 5: Vitamin C Content of Selected Vegetables and Fruits

(mg per 100 g serving)	
Red chili peppers	370
Guavas	240
Red sweet peppers	190
Parsley	172
Green sweet peppers	128
Broccoli	113
Brussels sprouts	102
Cauliflower	78
Red cabbage	61
Strawberries	59
Oranges	50
Lemons	46
Grapefruit	38

Source: Murray, M.T. (2001), *Encyclopedia of Nutritional Supplements: The Essential Guide for Improving Your Health Naturally*, New York: Three Rivers Press, p. 60. Reproduced with permission.

As Table 6 shows, the elderly and the malnourished are the most vulnerable to vitamin C deficiency. Since having coeliac disease predisposes a person to malnutrition, it is important that all patients with this disorder, and indeed anyone with a malabsorption problem, get themselves checked for levels of vitamin C.

Table 6: People Most Vulnerable to Vitamin C Deficiency

	% with low levels	*% with scurvy*
Young healthy	3	0
Elderly healthy	20	3
Elderly in hospital	68	20
Young institutionalised	100	30
Patients with cancer	76	46
Elderly institutionalised	95	50

Source: Murray, M.T. (2001), *Encyclopedia of Nutritional Supplements: The Essential Guide for Improving Your Health Naturally*, New York: Three Rivers Press, p. 61. Reproduced with permission.

The primary function of vitamin C is to manufacture collagen, which is the main protein in connective tissue, cartilage and tendons. Vitamin C is vital for the repair of wounds, keeping the gums healthy and preventing easy bruising.

However, vitamin C is also very important for immune function and this is probably the main reason why people use it as a supplement. This vitamin enhances the functions of white blood cells. It increases the level of interferon – the body's natural antiviral and anti-cancer substance. It also increases antibody levels and function, as well as having a positive effect on the thymus gland.

Using an analogy, if your white blood cells are soldiers then vitamin C is the entity that wakes them up and prepares them for action. There is sufficient scientific evidence to support the use of this vitamin in the treatment of various viral and bacterial infections, but there is an even more important role for vitamin C in *preventing* infections, especially in those prone to infections such as the elderly, the malnourished and those in hospital.

There is a lot of controversy regarding the recommended daily allowance or the RDA for all nutrients. The RDA is based on the minimum daily amount that will prevent the symptoms of deficiency; it is not the amount your body needs for optimal health. The RDA of vitamin C for humans is 60 mg, but the RDA for our close relatives in the zoo, monkeys and apes, is 2000 mg. I would suggest either making a fresh vegetable/fruit smoothie every day or taking a daily supplement of at least 1000 mg daily. In times of stress, pregnancy, surgery, trauma or infection, your daily intake of vitamin C should increase to a minimum of 1000 mg three times daily.

Since vitamin C is not stored in the body, it is not possible to overdose on it. It is very safe to use even in small children.

Pure vitamin C is called ascorbic acid; being an acid, it is best taken with food or water. It is also possible to buy it as calcium ascorbate or sodium ascorbate, which is less well absorbed by the body, but since it is not acidic it is easier to take. Personally I prefer the acid form as this is the natural form of the vitamin. Incidentally, it is the vitamin C that gives the fruits it is found in their acidic taste.

It's good to take vitamin C along with a good general vitamin/mineral supplement, as vitamin C needs other micronutrients to work effectively. There are many good vitamin/mineral supplements on the market. One of the best and one that has high levels of antioxidants is Male Multiple or Female Multiple by Solgar. The female version of this supplement contains iron but the male version does not, which is useful information for anyone with haemochromatosis (iron overload).

Zinc

A zinc deficiency can produce white spots on your nails – many people mistake this for a calcium deficiency – and constitutes one of the most common micronutrient deficiencies. This trace mineral is required as a co-factor in over 200 enzyme reactions in various organs of the body, so even a mild deficiency of zinc can have widespread effects on your health.

When I lived in southern Africa, where the soil is deficient in zinc, I saw the effects of low zinc on the body. Signs of a zinc deficiency include slow growth in children, poor appetite, mental lethargy and susceptibility to infections. If you or your child has a poor appetite or is prone to recurrent infections, a zinc supplement may be required.

Zinc is now firmly established as a major protector of the immune system and an important weapon in fighting disease. Zinc is involved in all aspects of immunity. When zinc levels are low, the number of T-cells reduces and many white blood cell functions decline. All of these effects are reversible if zinc is added to the diet.

Research on zinc deficiency has shown that there may be a decline in the number of lymphocytes in people over the age of 70. This section of the population is also more at risk of infection. It has been suggested that this weakening of the immune system as one gets older may be due to a fall in zinc levels. Other studies have shown that patients with AIDS have significantly lower blood levels of zinc compared with a control group. This suggests that there is a role for zinc supplementation in the elderly and in those with AIDS.

Here is a list of conditions that can be associated with a zinc deficiency:

- Alcoholism
- Anorexia
- Alopecia
- Connective tissue disease
- Delayed wound healing
- Delayed sexual maturation
- Dandruff
- Growth retardation
- Inflammatory bowel disease
- Infections
- Impotence and infertility
- Impaired glucose tolerance
- Loss of smell or taste
- Night blindness
- Skin disorders

The RDA for zinc is 15 mg for adults and 10 mg for children. I recommend a higher dosage, especially when you need to quickly break a cycle of recurrent infections. I normally recommend a zinc supplement for a three-month period and then reassess to see if it's necessary to continue.

You can also supplement your daily intake of zinc by adding certain foods to your diet. The best sources of zinc are wholegrains, legumes and animal meats, provided the soil in your area has adequate levels of zinc. Oysters have the highest level of zinc of all foods. Certain foods can affect the way your body is able to absorb and use zinc. Fibre, iron and calcium diminish the amount of zinc that you can absorb. Phytates in cereals can also bind to zinc and reduce its absorption.

There are no adverse effects associated with low-dose zinc supplementation. However, large doses of over 150 mg a day can have a negative effect on your immunity. For this reason correct dosage is important.

If zinc is used continuously for a period of time, it can create a copper deficiency; for this reason it is best to take zinc as part of a multi-mineral formulation that also contains copper, or, even better still, as part of a multi-vitamin/ mineral supplement.

Herbs

Echinacea – The Immunity Herb

Echinacea is famous for its ability to fight infection and boost the immune system. It is one of the most commonly used herbs in the world, particularly in the US and the alpine region of Europe. Echinacea is indigenous to North America, where it has been used for centuries by the Native American Indians for the treatment of infections and skin wounds. It was introduced to Europe by a doctor from Germany, Dr H.C.F. Mayer, who learned about its curative properties whilst on a visit to America. By 1907 it had become the most popular herb in medical practice in Germany. Today, it is used in herbal or homeopathic form in more than 250 medical products, including drops, tablets, capsules, ointments and creams.

It works by stimulating the white blood cells, which help fight infection. Research has shown that it enhances the activity of a particular type of white cell – macrophages. In December 1984, the medical journal *Infection and Immunity* reported that a particular ingredient of echinacea significantly increased the killing effect of macrophages on tumour cells (Stimpel *et al.*, 1984). Therefore, it has a role to play in the treatment of cancer as well.

Because echinacea can assist the body's defences, it helps in the control of viral, bacterial and fungal infections. It is also used to protect skin wounds.

Table 7 lists the conditions that may be helped by treatment with this herb.

Table 7: Uses of Echinacea

Infections	*Wounds*	*Other*
Colds and flu	Bites and stings	Cancer
Sore throats	Burns	Allergies
Ear infections	Skin ulcers	Eczema
Sinusitis		
Bronchitis		
Urinary infections		
Boils; abscesses		

Source: McKenna, J.E. (2003), *Natural Alternatives to Antibiotics*, Dublin: Newleaf, p. 56.

Echinacea is best taken as a liquid extract or tincture. In liquid form, it is more easily absorbed into the bloodstream, and as a medicine it will have a longer shelf life. I have found over the years that I get a much better result with an alcohol extract compared with the dried tablet or capsule form. The best liquid preparation is made by Bioforce and is sold under the tradename Echinaforce.

Echinacea can be used alone but in some cases is best combined with other herbs. For example, for bronchitis it can be combined with wild indigo, for immune enhancement it can be combined with Astragalus or wild indigo and for lymphatic drainage it can be combined with clivers.

Echinacea is one of the most researched herbs. It is also generally recognised as being one of the safest herbs: in all toxicity tests conducted by independent laboratories, it has been shown to be non-toxic. I personally have never known anyone to suffer from adverse effects as a result of taking this herb.

In summary, echinacea is one of the most important natural remedies for treating both acute and recurrent infections. It has a broad range of activity and is used to treat both internal infections and external or surface wounds. Its importance may increase dramatically as antibiotics become less effective in the years to come.

Chapter 9
Probiotics

At my public talks, I get asked about probiotics all the time. Probiotics represent the single most important nutritional supplement on the market. They are used in both children and adults and can treat a whole range of conditions.

If there is a class of 20 children and 7 of them get chickenpox and 13 don't, we tend to look at the children who contracted the infection. We say it is caused by a virus and we look at ways of killing the virus. But what about the children who didn't contract the virus? Shouldn't we look at them and try to figure out why they came off safe? If the virus (or any other microbe for that matter) is intrinsically harmful, it should have affected all 20 children. Maybe the 13 who escaped the infection have stronger immunity because of better constitutions, better diet, better microbial flora, etc. Therefore we should conclude that it may be possible to avoid infection by preventative measures.

However, prevention is not a money-making venture. There is money in finding drugs, antimicrobial drugs in this instance, and so Western society prefers to opt for this approach. It is failing miserably as it is based on a false premise: the belief that we can wipe out the bad bacteria and all will be well in the world. However, microbes are finding smarter ways to defend themselves.

This is where probiotics come in. Chapters 5 and 8 discuss how probiotics can be used to enhance the gut and the immune system. This chapter answers the most common questions I am asked about probiotics.

WHAT IS A GOOD PROBIOTIC?

There are a wide range of probiotics available in your local health shop or pharmacy. They come in many forms, such as capsules, powders, tablets, lozenges, chewing gum, drinks, foods and finally hydrotherapy implants.

In capsule probiotics, the capsule itself may be made of vegetable material or more commonly gelatin, or it may be enteric coated. Vegetable capsules are better as they have less moisture and keeping the bacteria dry is very important for their survival.

I notice on the labels of many modern probiotic capsule supplements that they may be left at room temperature. This makes sense as they have a very long shelf life when freeze dried and so can be stored safely at room temperature.

Probiotics in powder form are susceptible to damage because they are exposed to the air and, in particular, are not protected from moisture. Unlike capsules, they can deteriorate quite rapidly and become useless. If you need to use a powder, as is the case with children's formulations, then make sure the powder is kept dry by storing it in a dark container, sealing it and putting it in the fridge. Alternatively, you can buy the powder in sealed sachets, which is a much better option. Powders are best consumed with food such as dairy, e.g. yoghurt.

Tablets can be a better option, especially if the outer parts of the tablet can be used to protect the freeze-dried bacteria inside. However, capsules are easier to swallow and so are much more popular. Tablets also need to be kept dry and in the fridge.

Lozenges and chewing gum are new ways to introduce good bacteria into the mouth, ear, nose and throat areas of the body. They are especially useful in children prone to upper respiratory tract infections – ear infections and tonsillitis. It is an innovative way to encourage children to adopt the habit of taking a daily probiotic. Research has shown that using gum and lozenges can also help to reduce your chances of getting gum disease and dental caries (decay or cavity).

There are now many liquid probiotics in your local supermarket. I would avoid most of them as they contain sugar. If you can find one that is sugar free, does not contain artificial sweeteners and has been kept in a fridge, it can provide another innovative way of getting your children to take a daily probiotic.

There are also numerous probiotic foods in the marketplace, such as live yoghurt, and what I have said concerning liquids above applies here too. See Chapter 8 for more information on probiotic yoghurts.

Hydrotherapy inserts are another way of introducing good bacteria into the body. Hydrotherapy, also called colonic irrigation, is where a tube is inserted into the back passage and water is used to flush out the contents of the colon. It is an excellent therapy, especially if the therapist can insert good bacteria into the colon afterwards. There is, however, no scientific evidence that this way of using good bacteria is of clinical benefit.

However, if the colon is inflamed, as in cases of colitis or

Crohn's disease, then it is best to avoid colonic irrigation. This is because inflamed tissue can be quite fragile and so can be torn or perforated much more easily – if the colon is perforated this constitutes a medical emergency. If there is any evidence of colitis, Crohn's disease, diverticulitis, cancer of the colon, etc., then it is best to use an oral probiotic capsule.

DOSAGE

Dosage is measured in colony forming units (CFU). It is not a measure of the number of individual bacterial cells present but rather the number of bacterial colonies that appear when the powder is grown overnight on a growth medium in a laboratory.

For example, if 1 mg of powder produces 1 million CFU after incubation overnight, then it is estimated that 1 g of the powder will produce a 1,000 times that, i.e. 1 billion CFU.

I often get asked about the correct daily dosage of a probiotic. Scientific studies suggest a daily dosage of between 1 and 2 billion CFU, which is quite a high dosage. In truth, the correct dosage depends on two major factors. First, it depends on the species of bacteria present. With some species such as *Lactobacillus reuteri* a much lower dosage will suffice, as this bacterium is hardy and can colonise the wall of the stomach (a very acidic environment), duodenum and upper ileum. For example, it has been shown to inhibit the growth of the bacterium *Helicobacter pylori*, which causes peptic ulcers.

Second, the dosage will depend on your state of health. In general, the more ill you are the higher the dosage you need. In some cases, double or triple an average dosage will be needed for a short period of time. Also, if broad-spectrum antibiotics have been used, a higher dosage may be required.

Having said that, many commercial probiotics contain very low doses of bacteria per capsule, from 100 million to 1 billion CFU. This does not mean they are ineffective. Since bacteria multiply very frequently, this lower dosage may suffice if the supplement survives the stomach acid, if the supplement has some hardy species such as *L. reuteri* and if the patient's health is good. After all, some good bacteria are better than none.

The key factor to consider is getting the supplement past the stomach acid. Therefore, if you want the good bacteria to colonise the lower intestines, especially the colon, you must take the supplement when your stomach acid is at its lowest. This means taking a supplement first thing in the morning with a piece of fruit or some live yoghurt. Do not take it with a hot drink as high heat will kill most of the bacteria.

If you have been advised to use a probiotic twice daily, I would suggest taking it first thing in the morning and last thing at night. These are the two times of the day when the acidity of the stomach is at its lowest. Take it with a small amount of food, such as a piece of fruit, as the bacteria will need a little food to survive their trip down through the gut.

You can take a *prebiotic* along with the probiotic. As mentioned in Chapter 5, a prebiotic is food for the good bacteria. It can be taken separately but today many probiotic supplements encompass a prebiotic. It is labelled as FOS (fructooligosaccharide) or as inulin, as these are the most commonly used prebiotics. If using FOS or inulin separately, be careful with the dosage as using too much can cause bowel disturbances such as excessive flatulence. People with colitis, Crohn's disease or bowel cancer should avoid using FOS and inulin; rather, use a probiotic that does *not* have a prebiotic added.

In summary, a good probiotic dosage to aim for is around 1 billion CFU per day, but remember the dosage will depend on your level of health and the strain of bacteria in the supplement. Increase the dosage if you have recently taken an antibiotic or you are acutely ill. Take the probiotic with either a prebiotic or with a little food such as fruit or yoghurt.

HOW DO I KNOW WHICH PRODUCT TO BUY?

The simple answer is that it's quite hard to tell. A look at the manufacturer's website is a good place to start. You can tell how professional the company is from this. Also, look at the dosage on the packet. Is it labelled as CFU? If not, don't buy it unless recommended to do so by your practitioner.

Take a look at the bacteria included in the supplement. As a general rule it should contain *Lactobacillus acidophilus* and *Bifidobacterium bifidum*, often abbreviated to 'A' (for acidophilus) and 'B' (for bifidum), plus other species. So, make sure your supplement has A and B at least.

Many of the brands on the shelves today can be substandard. Either the formulation is not good, or they have not been manufactured or stored correctly. As a result, we have a whole range of good and not-so-good products, all adding to the consumer's confusion.

Studies carried out on probiotics for sale indicate that some brands fall far short of the claims made on the label. Research carried out at the University of Nebraska over a period of 20 years showed that 70 per cent of the products tested had fewer CFU than what was stated on the label; some had fewer than 10 per cent of the CFU claimed (Adams, 2009). This was confirmed by other studies.

The CFU count can drop in a supplement for a number of reasons, but by far the most common reason is that the product has not been stored correctly. Heat and water vapour will reduce the colony count very rapidly. It is therefore best to store all probiotic supplements in a sealed container in a cool, dry place.

FOR HOW LONG SHOULD I TAKE PROBIOTICS?

This a common question and much more difficult to answer. It depends entirely on the individual case. Obviously there is a huge difference between having an acute episode of diarrhoea, such as after the use of an antibiotic, and having a serious bowel disorder such as bowel cancer.

However, in general, for acute conditions, use a high dosage for two weeks and then a maintenance dose for a minimum of two months. For chronic conditions, use a high dosage for two months and then a maintenance dosage for two years. The exact dosage is best determined by your practitioner.

HOW DO I KNOW THE PROBIOTIC IS WORKING?

Again, this will vary with the condition of the person. However, in general, you should start to notice positive changes within two weeks. Your bowel motion should become firmer, meaning your stool should be better formed. It should be more solid and in one or two pieces.

Also the bowel habit should become regularised to once or twice a day. The bowel is supposed to empty after each significant meal and usually at the same time/s each day. In the Western world it is deemed good if one can achieve a once-a-day habit because of the low fibre content of the Western diet. It is good to take ground flaxseed as it bulks the stool as well as providing the body with essential oils.

Other changes in the gut may take a bit longer, such as a reduction in wind and bloating. Gradually, as the bowel habit improves and as the whole gut begins to function better, one's energy and overall health should improve. However, since there are trillions of bacteria in the body, it may take time for the probiotic to make an impact on such a large population. This is where persistence pays off. Use a good probiotic supplement for a sustained period of time.

WHAT I LOOK FOR IN A PROBIOTIC

I have answered some of the common questions I get asked about probiotics above. However, what really constitutes a good probiotic supplement?

First, the right supplement for you depends on your state of health. In other words, different strokes for different folks. For example, if you have been on one of the tetracycline antibiotics long term for the treatment of acne or to prevent malaria, I would choose to use 'For Those on Antibiotics' by OptiBac, whereas if you had chronic diarrhoea, I would opt for 'For Bowel Calm' also by OptiBac.

'For Those on Antibiotics' has two species of bacteria, *Lactobacillus rhamnosus* and *Lactobacillus acidophilus*, both of which have been proven to be of benefit in restoring the gut flora. 'For Bowel Calm' contains a completely different species called *Saccharomyces boulardii*, which has been shown in research studies to improve bowel habit.

There is no single supplement that suits everyone. The constituents of a chosen supplement must be of proven benefit to your clinical needs. The good thing about the company OptiBac is that their labels indicate what each supplement is suitable for, which means their formulations are well researched to achieve this aim.

The second point I would like to make is that it is not only the species of bacteria that is important but also the strain. Take the influenza virus that causes 'flu. Every year there are different strains of the same virus. So, even though you have immunity to one strain, this does not protect you from infection by other strains, as each strain behaves differently. Well, in the same way there are many strains of *Lactobacillus acidophilus*, each with different properties. Hence, it is very important to not only have the right species in your supplement but also the right strain.

On the supplement 'For Those on Antibiotics' is written the words *Lactobacillus acidophilus* Rosell-52. Rosell-52 is the strain. So, when choosing a supplement, always make sure the strain is included.

Third, there should be valid research data available from the company selling the product to show that the particular strain used in a formulation meets the following criteria:

- Ability to survive at room temperature
- Ability to survive stomach acid
- Ability to adhere to the wall of the gut and multiply
- Ability to counteract harmful bacteria

There should also be research information available on clinical trials that used the strain/s of bacteria in the formulation.

Fourth, the storage of the probiotics in your home is important. I was taught that heat and water were the two things to avoid when storing bacterial cultures. Nowadays, because most probiotic bacterial strains have been freeze dried, it is possible to store them at room temperature, so reducing the need to refrigerate them. However, they must still be kept dry.

The fifth and final point is related to the strength of the formulation. As there are trillions of bacteria colonising your body, you need at least billions in the formulation for it to have an impact. Aim for 1–2 billion CFU per daily dose. The strength of the formulation will decrease with time, so order your probiotic directly from the company if possible; otherwise, check the expiry date on the container or packet.

It is difficult to overestimate the importance of bacterial population to your survival. All living things are colonised with millions of bacteria. Your husband, wife or partner is composed mostly of bacteria – 90 per cent bacteria to be precise – your dog or cat is mostly bacteria, the cows and sheep in the fields are mostly bacteria, as are the plants that grow in your back garden. All living creatures need millions, billions and trillions of microbes to survive; without these microbes, life would not be possible.

Therefore, the most important contributors to your health and well-being are the bacteria that colonise your body. Look after these bacteria and they will take good care of you. This is why curds and whey, or a probiotic supplement, is the primary dietary supplement to focus on.

ALL BACTERIA ARE GOOD

A wise man said to me many years ago that everything in life is positive; it is only our conditioning that interprets some events in our lives as negative and others as positive. Having almost been killed during a house break-in, I came to appreciate what the wise man was saying.

While living in South Africa, my house was broken into in the middle of the night. I awoke with someone on top of me holding a gun to my head. Having grown up during the war in Northern Ireland, my greatest fear was to die violently; now, here in Africa, I was convinced my worst nightmare was about to happen. It didn't. Having ransacked the house and threatened myself and my daughter, they drove off in my car, which they had packed with most of my belongings.

I can honestly say that, for some time afterwards, I did not interpret this as a positive event. Having greeted death, I decided that the most important thing in my life was my family; everything else was less important. Even my work, which I had given a lot of importance to, took a back seat. I was grateful to be alive and able to share in the lives of my children.

The most important positive outcome of this event was that I was released from a deep-seated fear, the fear of violence. I was free to live a life without the controlling influence of this fear. I felt liberated and renewed, as if I had entered another phase of my life. So, in truth, the experience was both frightening and positive. I now began to understand the wise man a bit better. I also began to understand the old saying I had heard many times but never really understood: 'Your worst nightmare is your wildest dream trying to come true.'

Everything that happens to us in life can be seen as

positive, so all bacteria can be seen as positive too. The division of bacteria into bad and good is misleading. It distorts the truth about Nature and leads us into therapeutic cul-de-sacs.

This approach to bacteria is like dividing people into good and bad, and then killing off the bad to protect the good. The problem with this is that the definition of 'bad' would most likely be the result of shallow thinking, i.e. defining people on the basis of their actions, ignoring their underlying motivations. For example, if someone stole, they would be regarded as bad. I believe that nobody is intrinsically bad; people are intrinsically good. However, because of their environment or circumstances, such as poverty, they can end up committing an act that damages others. To rephrase, because of a change or trigger in their life (e.g. poverty), they become harmful to others.

In a similar way, *all* bacteria play a positive role in Nature. Otherwise they would not exist. However, if their environment alters, for example, when we consume a broad-spectrum antibiotic, one section of the bacterial population can be killed, allowing other opportunistic bacteria to overgrow. This is how most of the infections mentioned in earlier chapters, such as C.diff or E.coli, develop. These are not really harmful bacteria; they have become harmful because of a change in their environment.

Having lived in a divided society in Northern Ireland and then for many years in southern Africa, which was and still is divided by apartheid thinking – 'them' and 'us' – I can see that society is dominated by this thinking: one section of the community views the other as separate and in some way a threat. This mentality sets one against the other: Jew against Arab, Catholic against Protestant, black against white, Christian against Muslim, communist against capitalist, rich against poor. We preach unity but practise division. This has to be replaced by respect, not just for our fellow humans but for all living creatures, and that includes microbes.

To go back to the example I gave at the start of this chapter regarding the classroom of 20 children, 7 of whom contracted chickenpox. To me it makes much more sense to look at the 13 children in the class who did not get the infection and learn about the many effective ways of enhancing one's immunity, working closely with and not against Nature. This is the way to guarantee a safe future for humanity.

In summary then, we should try to view bacteria, viruses, fungi and other microscopic creatures in a more positive light and not continue with our approach of senseless extermination. Try to see bacteria as essential to your survival, not as some harmful alien from outer space hell bent on killing you. See them as your friends. They do, after all, constitute 90 per cent of your body.

Chapter 10

Allergic Reactions to Antibiotics

Antibiotics are commonly associated with allergic reactions. An allergic reaction can be related to the antibiotic itself or to the additives introduced during the manufacturing process. First, let's look at reactions to the drug itself.

REACTIONS TO THE DRUG

Between 5 and 15 per cent of people taking an antibiotic may experience a reaction. The penicillin group of antibiotics is widely regarded as most likely to cause an allergy, especially ampicillin and amoxycillin. These two antibiotics are most commonly prescribed for childhood infections.

It is not surprising, then, that most allergic reactions to antibiotics are seen in young children – those with tonsillitis, ear infections, chest infections, asthma, and urinary tract infections. As I have already discussed, most infections in children are viral, not bacterial, and do not require an antibiotic; those infections that are bacterial are best left for at least two days before starting treatment, to prevent the risk of a recurrence of the same infection.

However, other antibacterials can also cause an allergic reaction, such as the sulphonamides (tradenames Septrin and Bactrim) as well as the cephalosporins, the tetracylines and the infamous quinolones. In fact, any antibiotic has the potential to cause an allergic reaction, but by far the highest proportion of allergic reactions are due to penicillins.

The following case history details an allergic reaction to ampicillin.

Case History: Jane, Skin Rash

Jane was six years old and was put on ampicillin for a chest infection. She had a history of chest infections in winter but had never been on ampicillin as she was able to overcome them without medication. This infection was more serious as she had a fever and was coughing up greenish-coloured phlegm. After one week of being on the antibiotic, she developed a blotchy red rash on her trunk, which quickly spread to her limbs and face. The doctor suggested this may be an allergic reaction to ampicillin and stopped the drug. Within a few days, the rash cleared up completely.

This is an example of a mild allergic reaction to an antibiotic. Jane was told not to take ampicillin in the future. However, in cases like this, what I like to happen, though it seldom does, is that the person is given a medic alert bracelet to wear containing information about the allergy. I strongly urge you to get one if you have had a reaction to any drug. This would be really important if you happened to be admitted to hospital unconscious. It could save your life.

Hives or a nettle-sting-type rash is another symptom of a mild allergic reaction to a drug. Hives are caused by the release of histamine in the body and are characterised by red raised itchy spots with a pale centre. They can occur within 24 hours of being exposed to a drug. These hives can occur anywhere on the body but most typically on the trunk and limbs. They generally remain small and discreet, but if the person continues to take the drug the hives can enlarge and coalesce; blisters may also form in the mouth.

The next case history is an example of a severe skin reaction to penicillin, which can occur in a small percentage of patients. It is called Stevens–Johnson syndrome.

Case History: Alan, Tonsillitis

Alan was experiencing bouts of tonsillitis for a few months prior to seeking medical help. He went to his family doctor, who saw pus on his tonsils and immediately started him on a course of amoxycillin. Within days of being on the antibiotic, Alan developed a fever and felt quite a bit worse. He called his doctor, who advised continuing with treatment for a few more days. Then Alan began to develop skin lesions on his trunk and on the palms of his hands and soles of his feet. These lesions began to blister and blisters appeared on his eyes, nostrils and mouth.

By this stage Alan was in a lot more difficulty and had a cough and a bad headache. He spoke to his doctor again who immediately sent him to hospital where they diagnosed Stevens-Johnson syndrome, admitted him to the burns unit and stopped the drug. His skin began to peel off in sheets and tissue fluid began to ooze.

Fortunately, after one week of stopping the antibiotic, Alan's skin began to heal and, as his open lesions had not become infected, he was discharged and told not to take any penicillin drugs in future.

Alan can regard himself as very fortunate as this is a syndrome that can kill. It is a very severe form of allergic skin reaction to certain antibiotics. He was also lucky that the open wounds on his skin did not become infected, which is the real risk factor in cases like this. With good medical care in hospital, he was able to recover fully. Needless to say, he is very reluctant to take another antibiotic and prefers to use the alternatives, which is why he came to see me.

OTHER ALLERGIC REACTIONS

In addition to causing mild or severe skin reactions, antibiotics can affect other areas of the body such as the airway. This is particularly important for patients with an already existing respiratory disease.

Many asthma sufferers have had multiple courses of antibiotics; sufferers of chronic bronchitis and emphysema may also have had more than one course of an antibiotic, especially a broad-spectrum antibiotic. What they may not be aware of is that antibiotics can cause respiratory problems as part of an allergic reaction, as the next case history illustrates.

Case History: Brian, Asthma

Brian was only four years old when he was first diagnosed with asthma. He was diagnosed more on the basis of family history and a mild wheeze than any severe symptoms. On his fifth birthday he developed a bad cough and was bringing up yellow-coloured phlegm. The doctor prescribed an antibiotic called Penbritin (ampicillin). Within hours of taking it, Brian became quite short of breath and was gasping a lot. Initially his parents thought it was his asthma getting worse with the infection and gave him a puff of his Ventolin inhaler to help his breathing.

The next day, Brian had gotten worse in that he was still breathless but was also vomiting. His parents rang the doctor for advice. The doctor quickly recognised that Brian was having an allergic reaction to Penbritin and asked Brian's parents to stop the drug immediately.

Within hours of doing so, Brian had improved, and by the next day his breathing was back to normal and the vomiting had stopped. His phelgm had also become clear so the doctor held off prescribing another antibiotic. Within two days, Brian was back to normal.

This case history emphasises the need to always be on the alert for an allergic reaction when a penicillin antibiotic is being used. Ampicillin and amoxycillin can frequently cause skin, gut and airway reactions, as well as more severe reactions.

In truth, it can be very hard to detect an allergic reaction, especially if, as Brian's case illustrates, severe airway symptoms are present prior to taking the drug, such as cough, wheeze or shortness of breath. It is sometimes impossible to distinguish between the worsening of an underlying airway disease such as asthma and a drug allergy. If you have any doubt at all, it is best to stop the drug and seek medical advice.

An allergic reaction can be more dramatic than that described in Brian's case. It can cause severe shortness of breath and wheezing, making it very difficult for the person to breathe. It can require admission to hospital in some cases.

An allergic reaction may also be dominated by gut symptoms rather than those affecting the skin or respiratory tract. It may result in severe nausea or, as in Brian's case, vomiting. If you experience such symptoms soon after taking any drug, not just an antibiotic, stop the drug and see if the gut improves.

The worst type of allergic reaction to an antibiotic is anaphylactic shock, also called anaphylaxis. This is where a massive amount of the chemical histamine is released in the body, causing it to go into shock. As a consequence, it is a medical emergency.

For anaphylaxis to occur, the body usually has had a previous exposure to the drug, which caused a milder allergic reaction. On subsequent exposure, the body then reacts in a more dramatic way. This is why patients are advised not to take a drug that has previously caused even a very mild allergic reaction, as subsequent exposure may lead to anaphylaxis. This is also the reason why getting a medic alert bracelet is a good idea.

Because this severe reaction affects the whole body, there are many symptoms. Histamine inflames the blood vessels and causes leakage of fluid out of the blood vessels and into the tissues. Too little fluid in the blood vessels causes a drop in blood pressure and an increase in the pulse rate. Low blood pressure can make the person feel faint, dizzy or light headed, and it can also cause collapse. The fast pulse may cause palpitations.

Too much fluid in the tissues causes swelling and edema. It can also cause swelling of the face, tongue and eyes, resulting in slurred speech, difficulty swallowing and blurred vision. The skin may turn blue from a lack of oxygen.

The typical clinical picture is a history of exposure to a drug, swelling of the body, low blood pressure and fast heart rate. If these signs are present, urgent medical treatment is needed. This usually involves the administration of antihistamines, adrenaline and steroids. Other treatment may involve life-saving measures such as cardiopulmonary resuscitation (CPR), etc. But the single most important thing to do if someone is experiencing the symptoms of anaphylaxis is to call an ambulance and get the person to a clinic or hospital as soon as possible. Keep the patient warm and lying flat with their legs raised.

Although for the purposes of the present book we are discussing allergic reactions to antibiotics, anaphylactic shock can occur after exposure to almost anything your body reacts to, e.g. a bee sting, a food item such a peanut, pollen, etc. The body has the potential to react to anything it is exposed to, ingests or inhales.

REACTIONS TO THE ADDITIVES IN ANTIBIOTICS

It is also possible to have the full spectrum of allergic reactions to chemicals that are added to antibiotic drugs. These additives are mostly colourants and flavourings.

Colourants

Colourants are used to create an attractive colour. Red is the best colour to use in medicine as research shows it has the best compliance rate. In other words, people regard a red medicine as being stronger and so are more likely to follow instructions for taking the medicine and to complete the course.

The colourants used in antibiotics are mainly synthetic dyes. The two most common dyes used by the drug companies are tartrazine, which is yellow in colour, and Red Dye no. 40, which is red as its name suggests.

Tartrazine

Tartrazine is a dye that is used in many food items, including confectionery, drinks, vitamin tablets and drugs. It is also called Yellow Dye no. 5 and is known in Europe as E102. It gained a lot of publicity back in the 1990s when it was linked, along with many other colourants, to hyperactivity in children. It was then tested by researchers in the University of Southampton at the request of the Food Standards Agency in the UK. The results of this study, which were published in *The Lancet*, showed a clear link between hyperactivity in children when they consumed certain food additives, including tartrazine (McCann *et al.*, 2007). Because of this study, the UK government asked companies using tartrazine in their products to voluntarily remove it.

Tartrazine has a long history of causing behavioural changes in animals in experiments, and is also linked to ADHD (Attention Deficit Hyperactivity Disorder) in young children. A major study carried out by the University of Melbourne in 1994 showed evidence that consumption of tartrazine by children already diagnosed as hyperactive caused increased irritability, restlessness and sleep disturbances (Rowe and Rowe, 1994). It can also cause a range of allergic disorders, including skin itching, skin rashes and hives. This dye has also been linked with respiratory symptoms such as wheeziness and shortness of breath, and so may pose a threat to anyone with asthma. It can also cause gut symptoms such as nausea and vomiting, as well as abdominal cramps and pain.

Tartrazine is used by food and drug companies because it is very cheap. The pigment betacarotene is also yellow/orange in colour and would be a much safer option to use. Some countries in Europe have banned the use of tartrazine, but most have not. Norway banned the use of it but when Germany and Austria did likewise their national ban was overturned by the European Union, which still allows the use of tartrazine in EU member countries. It is still used widely in Canada and the US, where it is merely required that products containing the dye have accurate labelling.

There is sufficient experimental and clinical evidence to show that this dye can cause allergic reactions and changes in behaviour. It is in the best interests of humanity that this substance be removed from the food chain. However, until the general public object and make their voice heard, the interests of the food and drug companies will continue to come first.

Red Dye No. 40

Red Dye no. 40 is recognised as causing allergic reactions by the Food and Drug Administration (FDA) in the US. If it is ingested, it can cause mild skin reactions such as itching, rashes and hives, as well as more significant reactions, including gut symptoms such as nausea, abdominal cramps, wind and diarrhoea, and airway effects such as chest tightness, wheeziness and cough, itchy throat, sneezing and watery eyes. Even more serious reactions include tongue swelling, weakness and severe asthmatic symptoms through to anaphylaxis.

Flavourings – Aspartame

Flavourings are used to disguise the taste of the raw drug. Many antibiotics are either acidic or bitter and so have to be mixed with a more acceptable chemical or food item to make the final formulation more palatable. Sugar used to be the most commonly used flavouring, but today it has been replaced by artificial sweeteners. The artificial sweetener of choice for many pharmaceutical companies is aspartame. You may know it by one of its tradenames, NutraSweet or Equal. It is a popular sweetener, not only in pharmaceutical formulations (conventional drugs and antibiotics) but in many fizzy drinks and food items.

Not only can this chemical sweetener cause a range of allergic conditions, but it has been implicated in much more serious disorders. It replaced sugar back in the early 1980s and then quickly became the most popular sweetener, only to be surpassed by Splenda (sucralose) in the 2000s.

We were all convinced that the removal of sugar from fizzy drinks, many food items and medicines, and its replacement by an artificial sweetener, was a good thing, especially as sugar had been shown to cause obesity and a range of other disorders including heart disease. We could enjoy a soft drink without the worry of being exposed to too much sugar. A diet drink was the safe option. How wrong we were.

Aspartame is basically two amino acids joined together by methanol (poisonous liquid alcohol). When it is consumed, the gut breaks the chemical bonds holding the molecule together, releasing the methanol. Methanol is then converted in the body into formaldehyde. If you look up the US National Cancer Institute website, you will see formaldehyde listed as a carcinogen: www.cancer.gov/cancertopics/factsheet /risk/formaldehyde.

We now have both clinical and experimental evidence that this chemical causes blood cancers such as lymphoma, leukaemia and multiple myeloma (clinical tests involve humans and experimental tests involve animals). But evidence exists that it causes a whole host of other problems as well, including allergic reactions. Up to 10 per cent of people exposed to aspartame will exhibit an allergic response. The most common reaction is skin itching, which can be quite severe. Other people experience a skin eruption, including hives. In others, the airway or gut is affected.

The effect of aspartame on pilots has been the premise of a number of interesting studies. Much has been written about this, both in scientific literature and in pilot magazines. Dr Russell Blaylock, a neurosurgeon in the US, has studied this subject in detail. He blames the breakdown products of aspartame, which are known to affect the nervous system adversely, for symptoms experienced by pilots at high altitude, including disorientation, blurring of vision, difficulty thinking and concentrating, seizures, and heart failure. The symptoms are most acute if the pilot has skipped a meal and has low blood sugar levels.

Before its initial approval by the FDA for use in dry goods, in 1981, aspartame was not without its controversies. Aspartame was discovered by accident in a laboratory in Chicago in 1965 in the process of developing a new drug to treat ulcers. The chemist who discovered it worked for the pharmaceutical company G.D. Searle & Company. He realised that the substance he had made was very, very sweet – in fact 200 times sweeter than sugar. The company decided to do research on it over the coming years and applied for approval from the FDA. An application was made three times between 1975 and 1980, and each time it was rejected.

The rejections were due to the fact that aspartame was shown to have caused various tumours in laboratory animals (Stoddard, 1998), including brain tumours, pancreatic tumours, breast tumours and uterine tumours. Some of the by-products of aspartame metabolism were known to be carcinogenic.

In addition to rejecting aspartame, the FDA accused G.D. Searle & Company of scientific fraud and the case was referred to the Attorney General's office in Washington. This was because the FDA had proof that the data submitted to them by the company had been deliberately manipulated.

Many antibiotic preparations include suspect food additives that are considered safe by most food safety regulators worldwide. Despite evidence to the contrary, these regulators keep insisting there is no need to ban these additives. This suggests that regulators are not acting to safeguard the health of people.

In my opinion, if there is the slightest suggestion that a chemical can cause harm to anyone, it should be banned from our food and medication without the need for definitive evidence. Once banned, independent tests can be carried out to establish the toxicity of the additive. Human health is infinitely more important than the profit margins of food and drug companies.

When you are prescribed an antibiotic, your doctor is more interested in selecting the right antibiotic for your symptoms; for the most part, he or she ignores the food additives included in the capsule, tablet or syrup. If you have an allergic reaction, it is most likely your doctor will assume you are reacting to the drug.

The best person to discuss the ingredients of your medication with is your pharmacist. He or she will tell you about any food additives included in the formulation and advise you about safer alternatives.

Chapter 11

What Can We Do to Ensure the Prudent Use of Antibiotics?

The discovery of penicillin took medicine out of the dark ages and into the light. It allowed doctors to have a reasonable expectation of 'cure'. Because infections were now curable, this opened the way for greater survival rates post-surgery, in the intensive care unit (ICU), after childbirth and after major accidents. It also opened the way for chemotherapy to help increase survival rates in cancer patients. Chemotherapy wipes out the immune system, making the person very vulnerable to infection. Many cancer patients die from infections caused by the disease itself or by the therapy.

It is difficult to fully appreciate the profound effect that antibiotics have had on every branch of medicine and on society as a whole. They facilitated expansion in many areas of medicine and allowed for the development of invasive procedures such as laparoscopy, angiograms, open heart surgery, organ transplantation and even the use of simple catheters. Modern medicine is totally dependent on antibiotics. Without them it would collapse.

It is therefore imperative that our medical advances be protected. To do this we must lobby for the prudent use of these precious drugs. As you saw in Chapter 2, the only doctors with reasonable knowledge regarding the use of these drugs are infectious disease experts (in Europe we use the term microbiologists). I believe that other doctors should not be allowed to prescribe antibiotics unless they can prove through valid tests that the infection is bacterial. The valid test result proving an infection to be bacterial would then be faxed or emailed to the pharmacist as a necessary requirement for the dispensing of antibiotics. The prescribing and dispensing of antibiotics could be controlled by infectious disease experts, who could withdraw an antibiotic for a period of time to allow bacteria to lose their resistance gene and become sensitive to that antibiotic again.

Alternatively, all prescriptions for antibiotics could be approved by an infectious disease expert before being dispensed by a pharmacist. In this way, the expert would have complete control over the use of these drugs, which would guarantee their continued worth.

Of course, this only addresses the issue of medical prescriptions and does not deal with the widespread use of these drugs in farming and in countries where they are available over the counter or on the black market. It falls on veterinary schools to control all prescriptions from vets and control what farmers are allowed to do in each country. Serious fines should be given to those caught selling antibiotics without a prescription. Countries that fail to bring this problem under control should face heavy penalties.

As recently as April 2014, the World Health Organization (WHO) issued a warning that if care is not taken in the use of antibiotics we will face a post-antibiotic era where antibiotics will be ineffective at treating infection (Boseley, 2014). Curtailing the use of antibiotics is one of the biggest challenges facing the human population. It does not require more seminars, conferences, meetings and words of warning. We have passed the point of wise words and now require action. Action is really the domain of the WHO, but unfortunately this organisation has proven itself to be subject to political interference (McKenna, 2013).

It will require the government of each individual country to urgently implement measures to control the prescribing, dispensing and sale of these drugs. If each government acts now, there is a reason to be optimistic. If they do not, all the warnings of doom from the experts may well be realised.

WHAT YOU CAN DO

The first and perhaps the simplest measure to prevent the spread of infection is to maintain good hygiene. Washing your hands with soap and water after a visit to the toilet or after handling raw foods, especially meats, is an extremely effective means of reducing the spread of infection within the family and within the community. You could be carrying MRSA, for example, and not even be aware of it.

It is very important for farmers to wash their hands, as they, especially those who have livestock, carry bacteria with the highest levels of resistance. This is because livestock are constantly being exposed to antibiotics, as antibiotics are still available in animal feed in many countries (excluding EU member countries), and are also available over the counter from veterinary surgeons in all countries. Exposure to many different antibiotics means that the bacteria that colonise the animals develop multiple drug resistance. These bacteria are then spread to anyone who handles the animals, including farmers and veterinary surgeons. Therefore, hand washing is really important.

A recent study from Michigan State University in the US found that only 5 per cent of people wash their hands long enough with soap and water to effectively kill all the bacteria on their hands. Of the almost four thousand people observed, one third did not use soap (Borchgrevink *et al.*, 2013). However, soap is necessary to kill bugs. It takes a mere 20 seconds of vigorous hand washing with soap and water to produce bug-free hands. It is worth it if we can prevent the deaths of people in our family, village or town.

The second important thing that you can do is wait; don't rush to your doctor for an antibiotic at the first sign of an infection. Wait to see if your body can fight off the infection, as early use of an antibiotic encourages the recurrence of the infection and also encourages bacterial resistance. However, if you are not able to fight the infection, ask your doctor to take a swab, urine sample or stool sample, etc. and send it to the laboratory. This will indicate if the infection is bacterial and what antibiotics will be effective.

As you saw in Chapter 2, there are major misconceptions among the public about when an antibiotic is appropriate and when it is not. To tackle these misconceptions, the Centers for Disease Control and Prevention (CDC) in the US have been running a campaign called 'Get Smart: Know When Antibiotics Work' to educate parents and adults about the use of these drugs.

There have been similar campaigns in Europe to try to

control the use of these drugs in the community. For example, the campaign in France called 'Les antibiotiques, c'est pas automatique' (Antibiotics are not automatic) has been very effective. It was quite a comprehensive national campaign, involving posters, exhibitions, the internet and social media. It began in 2002 and after five years there was a 25 per cent drop in antibiotic prescriptions. It appears to be working and is still ongoing today (WHO, 2011).

The wonderful Europe-wide website www.e-bug.eu is a free resource for school teachers and for school children of all ages. The aim of this interactive website is to teach children about hygiene and about microbes, and when to treat an infection and when not to. It has a number of games that children can play and also has a number of competitions for older students. Check it out and see what you think.

The third thing we can do is educate ourselves about the misuse of these drugs, and educate our children, our friends and relatives and, if possible, our community. If you see an antibiotic being used in a hospital, by your family doctor or by your veterinary surgeon, ask if it is entirely necessary. Ask if there are other ways of treating the infection.

Your participation is really important if your children and future generations are to avoid death from common infections. Today, since we still have some ability to help people with common infections, it may not seem urgent that you do anything. 'The authorities will solve the problem' is a refrain I hear quite often. If catastrophe is to be avoided, it requires action from everyone.

The fourth thing that you can do is protect yourself from getting an infection in the first place. This means eating well and drinking lots of water. Avoid sugary foods and sugary drinks, and avoid white flour. Eat lots of natural produce such as eggs, cheese, butter, wholegrains, red and white meats, and organ meats, especially liver. For more detail, read my book *Good Food* (2013).

The best way to avoid an infection is to include curds and whey in your diet, such as live yoghurt, and to take a daily probiotic, such as 'For Every Day' by OptiBac.

WHAT DOCTORS CAN DO

First, doctors can give their prescription pads a rest. By adopting a wait-and-see approach to most infections, they will do the whole community a service. It is wiser to wait for 48 hours to see if the body can fight an infection or not. In this 48-hour period, laboratory tests can confirm the presence (or absence) of a bacterial infection and show what antibiotic to use. The approach of prescribing antibiotics 'to be on the safe side' must be reviewed very seriously. Doctors examining a young child with a suspected infection in Accident and Emergency (A&E) will very often cover the possibility of a bacterial infection by prescribing a broad-spectrum antibiotic. Family doctors also have to guess and very often prescribe when they are in doubt. This is bad medicine as it is compromising us all. The old proverb 'When in doubt, do without' applies here.

Second, doctors must become more aware of the unhealthy relationship between medicine and drug companies. There are far too many cases of pressure being applied to doctors by drug reps, who encourage them to prescribe more and more. Drug companies are interested in profits, not a patient's health. Doctors should divorce themselves from drug companies so that they can offer objective advice. As I discussed above, I feel that if the problem of resistance is to be taken seriously, doctors should not be allowed to prescribe an antibiotic without the consent of an infectious disease consultant.

Third, doctors working in rural dispensing practices must have their prescribing patterns monitored more closely. As discussed in Chapter 6, there is evidence now to suggest a link between diabetes and the over-prescription of antibiotics by dispensing practices.

Fourth, doctors must inform their patients about how best to prevent infections in the first place and what natural options are available via their pharmacy or health shop.

WHAT FARMERS CAN DO

Farmers must lobby for the removal of all antibiotics from animal foodstuffs, if the country they live in has not banned them already. Eighty per cent of all the antibiotics produced worldwide are used as growth promoters, either in animal feed or via injections. This is clearly a practice that is doing more harm than good. It promotes the growth of multi-resistant superbugs in the animals, which can be contracted by the farmer and his family, and can ultimately affect the whole community. This is why the EU banned the use of all antibiotics in animal feed back in 2007.

However, antibiotics are still widely available to farmers to treat infections in their livestock. In many cases, the farmer can buy the antibiotic without the veterinary surgeon having examined the animal. Sometimes the antibiotic is available over the counter in veterinary practices without the knowledge of the veterinary surgeon. Nowhere in the community should it be possible for an antibacterial drug to be sold without a prescription, regardless of whether it's for human or animal use.

I have seen farmers give animals antibiotics for a sprain. There clearly needs to be an educational campaign directed at farmers and their families so that this inappropriate use of these drugs ceases. An awareness programme for farmers, along the lines of the French campaign 'Les antibiotiques, c'est pas automatiques', could be appropriate.

Farmers and their families are the ones most at risk of contracting an infection caused by a superbug so it is in their own best interests to address this issue. Farmers also have a powerful political lobby via their national associations, which they could harness to effect change within their country. If farmers became more proactive, they would guarantee a safer future for all of us.

WHAT VETERINARY SURGEONS CAN DO

As with doctors, veterinary surgeons must also hold back from prescribing antibacterial drugs too soon. They should opt to wait and see how the infection progresses, and to use natural treatments in the meantime. They can also use this time interval to send a swab/blood/urine/stool sample to the laboratory for analysis. Veterinary surgeons usually dispense these drugs as well as prescribe them. This is a practice that needs to be reviewed by the relevant authorities, since we are facing a potentially serious situation.

Because of their daily contact with animals, veterinary workers, like farmers, are more at risk of carrying multi-resistant bacteria and so need to scrub their hands well between examining each animal. They should carry out a surgical scrub at the end of their working day.

It may well prove fruitful for a good educational programme to be targeted at all veterinary practices within a country, if this has not already been done. This could be organised by the major veterinary colleges or be combined with a programme directed at farmers.

WHAT THE DRUG COMPANIES CAN DO

First and foremost, the drug companies need to act more responsibly in terms of pressurising doctors into prescribing more antibiotics. Drug reps need to be made aware of the dangers associated with the use and misuse of these drugs.

In addition, the drug companies need to accept more responsibility for the discovery and development of new antibacterial drugs. Certain drug companies have gone some way along this road. These companies have favoured the development of synthetic drugs, which are manufactured in laboratories. Initially, it was thought that bacteria would be less likely to develop resistance to these drugs so rapidly. Also, synthetic drugs are easier to manufacture and are not contaminated with unknown substances. However, after years of endeavour in this direction, there are no new synthetic antibacterials.

Now it seems these companies may have to go back to basics and examine fungi and some other microbes that are known to produce antibacterials in Nature. These companies have also begun to realise that it is smarter to work with Nature and adopt a more long-term approach, so that future generations can also benefit from the amazing power of antibiotics and be spared the huge problems associated with superbugs. There seems to be a light at the end of the tunnel – maybe!

Drug companies must work with governments to establish private–public partnerships to facilitate the research and development of new drugs, to control the use of any future antibiotics to minimise resistance, and to educate everyone in the community about the responsible use of these drugs.

As I mentioned at the start of this book, many drug companies have abandoned the search for new antibacterial drugs because it takes too long to get the drug approved, it is a very costly exercise – up to $1 billion per drug – and bacteria can render the drug ineffective in a short space of time. No new class of antibiotic has been developed since 1987. Governments need to address these issues and encourage the search for and development of new antibiotics. More money could also be given to universities to facilitate this search.

WHAT GOVERNMENTS CAN DO

Governments can facilitate the development of new diagnostic techniques so that the diagnosis of a bacterial infection can be made much earlier. It is now possible to use DNA finger-printing techniques, which will not only identify if the microbe is bacterial but will also indicate the species and strain. Current procedures involve a sample being sent to the laboratory, where the organisms are incubated overnight and then examined under the microscope. This can take time. DNA finger-printing is much faster. It may be possible to develop a kit that can be used on the ward or in a doctor's office.

Governments have to give the prudent use of antibiotics top priority and allocate adequate funding to ensure that the actions mentioned in this chapter are instigated. As mentioned, Professor Sally Davies, Chief Medical Officer for England, has linked the threat of antibiotic resistance to that of terrorism. Urgent action is required.

In the past, as with the issue of global warming, governments have been very slow to act. It is up to us, the general public, to lobby our politicians and push for effective measures. Just pushing for strict controls on the dispensing and sale of these drugs will help. Do what you can to help solve this problem if you wish to see modern medical care survive.

Chapter 12
Summary

Antibiotics are big business. Most antibiotics are prescribed for paediatric problems such as ear and throat infections. In the US alone, over half a billion dollars' worth of antibiotics are sold each year for one infection only – middle ear infection (*otitis media*). If we take the range of infections in all countries into account, it is safe to say that antibiotics are a multi-billion dollar industry.

Children are bearing the brunt of the side effects of these drugs. Further, the over-prescription of antibiotics to children is making them more vulnerable to multi-resistant bacteria. It is time to protect the vulnerable in society by condemning the senseless misuse of these drugs and calling for a saner, more intelligent approach to their use. It is time to cast a bright light on the relationship between the medical profession and the drug companies, and to cast an even brighter light on the relationship between politicians and the drug companies.

Because of the misuse of these drugs by Western medicine, by the manufacturers of livestock feed and by the veterinary profession, we now have a situation where bacteria have outsmarted everyone and rendered our magic bullets (antibiotics) almost useless. As time marches on, more and more bacteria are becoming resistant to more and more antibiotics, allowing for the emergence of superbugs.

Superbugs used to be limited to hospitals; today they are in the community. Common infections are proving more and more difficult to treat. Soon we will regress to pre-penicillin days, when people died from ear infections, sore throats, abscesses, etc. Because so much of modern health care depends on the use of antibiotics, many routine procedures such as endoscopy and common operations such as hip replacements will become too risky.

We are in this mess for a number of reasons. First, we have misused these drugs and continue to misuse them, despite all the public warnings not to do so. The case histories in this book are a testimony to this.

We have grossly underestimated the so-called enemy, pathogenic bacteria. They have been around on the planet for a lot longer than humans and have learned many innovative survival tactics, such as creating resistance genes and sharing these genes with other bacteria.

We have also underestimated bacteria by assuming that if we kill enough of them we will win the battle. Bacteria on average multiply every 20–30 minutes. So, even if you killed all bar one, twelve hours later you would have billions of them to deal with all over again.

We have made the gross error of not advising patients to take a probiotic at the same time as using an antibiotic. This error has led to a whole host of opportunistic infections, such as C.diff, taking hold.

The argument that we need more antibiotics urgently is valid, but if bad habits are not corrected – the same bad habits that caused this mess – then new antibiotics will not solve the problem.

The core error we made was to view some bacteria as 'bad' and to set about exterminating them from the planet. This has not only failed miserably but has landed the human population in difficulty. As I have tried to put across in this book, it is best to view all forms of life and planet Earth itself with respect. This simplistic approach of viewing things as either good or bad is harming us all.

If these errors in our approach to antibiotics are corrected then there will indeed be hope for our children and grandchildren. This will require a paradigm shift of enormous proportions. It essentially requires a more spiritual approach to our view of ourselves and the world we occupy. Therefore, if we really want things to improve, we each must become more human, more understanding, more sympathetic and more empathetic towards each other and other living beings. This I see as the ultimate solution.

Finally, if there is one message I've tried to get across in this book, it is this: *treat antibiotics with great respect.*

John McKenna
E-mail mckennaje@hotmail.co.uk

Appendix: Commonly Used Antibiotics

These are some of the most commonly used antibiotics, but they will be replaced in years to come as new drugs come on the market, or as the present drugs become ineffective and have to be withheld or discontinued. Tradenames may vary from country to country, especially between the countries of Europe and those of North America.

Penicillins

Antibiotic	*Tradename*
Amoxycillin	Amoxil or Novamox
Ampicillin	Penbritin or Omnipen
Flucloxacillin	Floxapen
Penicillin	Pfizerpen
Ticarcillin	Ticar

Penicillin Combinations

Antibiotic	*Tradename*
Amoxycillin-Clavulanate	Augmentin
Ampicillin-Sulbactam	Unasyn
Piperacillin-Tazobactam	Zosyn

Tetracyclines

Antibiotic	*Tradename*
Minocycline	Minocin
Doxycycline	Vibramycin
Oxytetracycline	Sumycin or Achromycin
Demeclocycline	Declomycin

Cephalosporins

Antibiotic	*Tradename*
Cefadroxil	Duricef
Cefazolin	Ancef
Cefalexin	Keflex
Cefaclor	Distaclor
Cefuroxime	Zinnat
Cefixime	Suprax
Cefdinir	Omnicef
Cefditoren	Spectracef
Ceftazidime	Fortaz

Aminoplycosides

Antibiotic	*Tradename*
Amikacin	Amikin
Gentamycin	Garamycin
Kanamycin	Kantrex
Neomycin	Neo-Fradin
Tobramycin	Nebcin

Anti-Tuberculosis Drugs

Antibiotic	*Tradename*
Cycloserine	Seromycin
Ethambutol	Ethambutol
Isoniazid	Isoniazid
Pyrizinamide	Pyrizinamide
Rifampicin	Rifadin or Rimactane or Rifinah
Streptomycin	Streptomycin

Quinolones

Antibiotic	*Tradename*
Ciprofloxacin	Ciproxin, Ciprobay
Levofloxacin	Tavanic
Moxifloxacin	Avelox
Nalidixic Acid	Mictral or Negram
Ofloxacin	Tarivid, Floxin, Ocuflox
Gatifloxacin	Tequin

Others

Antibiotic	*Tradename*
Erythromycin	Erythroped or Erymax
Azithromycin	Zithromax
Co-trimoxazole	Septrin
Trimethoprim	Monotrim or Trimopan
Metronidazole	Flagyl

Bibliography

Adams, C. (2009), *Probiotics: Protection Against Infection*, Delaware: Sacred Earth Publishing.

Avila, J. (2012), 'Superbug dangers in chicken linked to 8 million at-risk women', ABC News, 11 July, available from: http://abcnews.go.com.

Bekelman, J.E., Li, Y. and Gross, C.P. (2003), 'Scope and impact of financial conflicts of interest in biomedical research: a systemic review', *Journal of the American Medical Association*, vol. 289, no. 4, pp. 454–65.

Blumenthal, D. (2004), 'Doctors and drug companies', *New England Journal of Medicine*, vol. 351, no. 18, pp. 1855–90.

Borchgrevink C.P., Cha, JaeMin and Kim, SeungHyun (2013), 'Hand Washing Practices in a College Town Environment', *Journal of Environmental Health*, vol. 75, no. 8, pp. 18–24.

Boseley, Sarah (2014), 'WHO calls for urgent action to preserve power of antibiotics and make new ones', *The Guardian*, 30 April.

Bratu, S., Landman, D. and Haag, R. *et al.* (2005), 'Rapid spread of carbapenem resistant *Klebsiella pneumoniae* in New York City: a new threat to our antibiotic armentarium', *Archives of Internal Medicine*, vol. 165, no. 12, pp. 1430–5.

Campbell, E.G., Weissman, J.S. and Ehringhaus, S. *et al.* (2007), 'Institutional Academic-Industry Relationships', *Journal of the American Medical Association*, vol. 298, no. 15, pp. 1779–86.

Cantekin, E.I., Maguire, T.W. and Griffith, T.L. (1991), 'Antimicrobial therapy for *otitis media* with effusion', *Journal of the American Medical Association*, vol. 266, no. 23, pp. 3309–17.

Cardwell, C.R., Carson, D.J. and Patterson, C.C. *et al.* (2006), 'Higher incidence of childhood-onset type 1 diabetes mellitus in remote areas: a UK regional small area analysis', *Diabetalogia*, vol. 49, no. 9, pp. 2074–7.

Centers for Disease Control and Prevention (2012), 'Vital Signs: Preventing *Clostridium difficile* Infections', *Morbidity and Mortality Weekly Report*, 9 March, vol. 61, no. 9, pp. 157–62.

Chambers, H.F. (2001), 'The changing epidemiology of *Staphylococcus aureus*?' *Emerging Infectious Diseases*, no. 7, 178–82.

Chandler, D. and Dugdale, A.E. (1976), 'What do patients know about antibiotics?' *The Lancet*, vol. 308, no. 7892, p. 422.

Coutsoudis, A., Bobat, R.A. and Coovadia, H.M. (1995), 'The effects of vitamin A supplementation on the morbidity of children born to HIV infected mothers', *American Journal of Public Health*, vol. 85, no. 8 (Pt 1), pp. 1076–81.

Davies, Sally (2013), *The Drugs Don't Work: A Global Threat*, London: Penguin.

Delia, A., Morgante, G. and Rago, G. *et al.* (2006), 'Effeciveness of oral administration of *Lactobacillus paracasei* in association with vaginal suppositories of *Lactobacillus acidophilus* in the treatment of vaginosis and in the prevention of recurrent vaginitis', *Minerva Ginecol.*, vol. 58, no. 3, pp. 227–31.

Diamant, M. and Diamant, B. *et al.* (1974), 'Abuse and timing of the use of antibiotics in acute *otitis media*', *Archives of Otolaryngology*, vol. 100, no. 3, pp. 226–32.

Drago, L., De Vecchi, E. and Nicola, L. *et al.* (2007), 'Activity of a *Lactobacillus acidophilus*-based douche for the treatment of bacterial vaginosis', *Journal of Alternative and Complementary Medicine*, vol. 13, no. 4, pp. 435–8.

Falkow, S. and Kennedy, D. (2001), 'Antibiotics, Animals and people – Again!' *Science*, vol. 291, no. 5503, p. 397.

Fleming, A. (1945), 'Penicillin', Nobel lecture, 11 December, available from: www.nobelprize.org.

Hauser, W.E. and Remington, J.S. (1982), 'Effects of antibiotics on the immune response', *American Journal of Medicine*, vol. 72, no. 5, p. 711–16.

Health Canada (2003), 'Canadian Adverse Reaction Newsletter', July, vol. 23, no. 3, available from: www.hc-sc.gc.ca.

Hlivak, P., Jahnova, E. and Odraska, J. *et al.* (2005), 'Long-term oral administration of the probiotic *Enterococcus faecium* M-74 decreases the expression of sICAM-1 and monocyte CD54 and increases that of lymphocyte CD49d in humans', *Bratislav. Lek. Listy*, vol. 106, no. 4–5, pp. 175–81.

Jaffe, D.M., Tanz, R.R. and Davis, A.T. (1987), 'Antibiotics Administered to treat possible occult bacteremia in febrile children', *New England Journal of Medicine*, vol. 317, pp. 1175–80.

Jarvis, W.R. (2013), *Bennett & Brachman's Hospital Infections*, 6th edition, Philadelphia: Lippincott-Raven.

Kaye, Donald (2008), 'News: TB exposure found on India-to-US flight', *Clinical Infectious Diseases*, vol. 46, no. 5, p. iii.

Kloppenburg, M., Verweij, C.L. and Miltenburg, A.M. *et al.* (1995), 'The Influence of tetracyclines on T-cell activation', *Clinical & Experimental Immunology*, vol. 102, no. 3, pp. 635–41.

Kuntaman, K., Lestari, E.S. and Severin, J.A. (2005), 'Fluoroquinolone resistant E.coli, Indonesia', *Emerging Infectious Diseases*, vol. 11, no. 9, pp. 1363–9.

Landymore-Lim, L. (1994), *Poisonous Prescriptions; Antibiotics Can Cause Asthma*, Subiaco, Australia: PODD.

Lazarow, A. *et al.* (1951), 'Protection against alloxan diabetes with cobalt, zinc and ferrous iron', *The Anatomical Record*, vol. 109, p. 377.

Lidefelt, K.J., Bollgren, I. and Nord, C.E. *et al.* (1991), 'Changes in periurethral microflora after antimicrobial drugs', *Archives of Disease in Childhood*, vol. 66, no. 6, pp. 683–5.

Llor, C. and Cots, J.M. (2009), 'The sale of antibiotics without prescription in pharmacies in Catalonia, Spain', *Clinical Infectious Diseases*, vol. 48, no. 10, pp. 1345–9.

McCann, D., Barrett, A. and Cooper, A. *et al.* (2007), 'Food additives and hyperactive behaviour in 3 year old and 8/9 year old children in the community: a randomised double blind placebo controlled trial', *The Lancet*, vol. 370, no. 9598, pp. 1560–7.

McGreal, C. (2007), 'Nigeria sues Pfizer for £3.5bn over illegal child drug trial', *The Guardian*, 6 June, available from: www.theguardian.com/business/2007/jun/06/medicineandhealth.health.

McKenna, J.E. (2013), *Good Food: Can You Trust What You Are Eating?* Dublin: Gill & Macmillan.

McKenna, J.E. (2002), *Hard to Stomach: Real Solutions to Your Digestive Problems*, Dublin: Newleaf.

McKenna, J.E. (2003), *Natural Alternatives to Antibiotics*, Dublin: Newleaf.

Mandel E.M., Rockette, H.E., Bluestone, C.D. *et al.* (1987), 'Efficacy of Amoxycillin with and without decongestant-antihistamine for *otitis media* with effusion in children', *New England Journal of Medicine*, vol. 316, no. 8, pp. 432–7.

Metcalfe, M.A. and Baum, J.D. (1991), 'Incidence of insulin dependent diabetes in children aged under 15 in the British Isles during 1988', *British Medical Journal*, vol. 302, no. 6774, pp. 443–7.

Murray, M.T. (2001), *Encyclopedia of Nutritional Supplements: The Essential Guide for Improving Your Health Naturally*, New York: Three Rivers Press.

Niebergall, P.J., Huzzar, D.A. and Cressman, W.A. *et al.* (1966), 'Metal Binding Tendencies of Various Antibiotics', *Journal of Pharmacy & Pharmacology*, vol. 18, no. 11, pp. 729–38.

Pichichero, M.E., Disney, F.A. and Talpey, W.B. (1987), 'Adverse and beneficial effects of immediate treatment of Group A Beta-hemolytic streptococcal pharyngitis with penicillin', *Paediatric Infectious Diseases Journal*, vol. 6, pp. 635–43.

Poses, R.M., Cebul, R.D. and Collins, M. *et al.* (1985), 'Assessment of the diagnostic accuracy of experienced physicians', *Journal of the American Medical Association*, vol. 254, no. 7, p. 927.

Rennie, D., (1991), 'The Cantekin Affair', *Journal of the American Medical Association*, vol. 266, no. 23, pp. 3333–7.

Reusser, Fritz (1971), 'Mode of Action of Streptozotocin', *Journal of Bacteriology*, vol. 105, no. 2, pp. 580–8.

Rowe, K.S. and Rowe, K.J. (1994), 'Synthetic food colouring and behaviour: a dose response effect in a double-blind, placebo-controlled, repeated-measures study', *Journal of Paediatrics*, vol. 125, no. 5 (Pt 1), pp. 691–8.

Sanchez, A., Reeser, J.L. and Lau, H.S. (1973), 'Role of sugar in human neutrophilic phagocytosis', *American Journal of Clinical Nutrition*, vol. 26, no. 11, pp. 1180–4.

Schmidt, M.A. (1993), *Beyond Antibiotics: 50 (or So) Ways to Boost Immunity and Avoid Antibiotics*, US: North Atlantic Books.

Spellberg, B. (2008), 'William Stewart, MD: mistaken or maligned?' *Clinical Infectious Diseases*, vol. 47, no. 2, p. 294.

Stimpel, M., Proksch A. and Wagner H. (1984), 'Macrophage activation and induction by purified polysaccharide fractions from the plant *Echinacea purpurea*', *Infection & Immunity*, vol. 46, no. 3, pp. 845–9.

Stoddard, M.N. (1998), *Deadly Deception: The Story of Aspartame*, Dallas: Oderwald Press.

Sugawara, G., Nagino, M. and Nishio, H. (2006), 'Perioperative symbiotic treatment to prevent postoperative infectious complications in biliary cancer surgery: a randomized, controlled trial', *Annals of Surgery*, vol. 244, no. 5, 706–14.

Tadros, W.M. *et al.* (1982), 'Protective effects of trace elements on alloxan-induced diabetes', *Indian Journal of Experimental Biology*, vol. 20, pp. 93–4.

Thomas, Lewis (1995), *The Youngest Science: Notes of a Medicine-Watcher*, London: Penguin.

Thong, Y.H. and Ferrante, A. (1980), 'Effect of tetracycline treatment on immunological responses in mice', *Clinical & Experimental Immunology*, vol. 39, no. 3, pp. 728–32.

Walsh, F. (2013), 'Antibiotics resistance "as big a risk as terrorism" – medical chief', BBC News, 11 March, available from: www.bbc.com/news/health-21737844

Werbach, M.R. (1996), *Nutritional Influences on Illness: A Sourcebook of Clinical Research*, 2nd edition, California: Third Line Press.

WHO (2011), 'Are antibiotics still "automatic" in France?' Bulletin of the World Health Organization, vol. 89, no. 1, available from: www.who.int/en.

WHO (2014), 'Tuberculosis', Factsheet no. 104, available from: www.who.int/mediacentre/factsheets/fs104/en/.

Wong, T. and Tiessen, E. (1989), 'Managing sore throat in theory versus practice', *Canadian Family Physician*, vol. 35, pp. 1771–3.

Resources

SUPPLIERS OF GOOD WATER

Isklar, Sabco Group, PO 3779, Postal Code 112, Ruwi, Sultanate of Oman. www.sabcogroup.com. Tel. 968 2466 0100. E-mail info@sabcogroup.com

Spa, Brecon Water, Trap, near Llandeilo, Carmarthenshire, Wales SA19, 67T. www.breconwater.co.uk. Tel. 44 1269 850175. E-mail sales@breconwater.co.uk

Pellegrino S.p.A., via Lodovico Il Moro 35, 20143 Milan, Italy. www.sanpellegrino.com

Volvic Water, Danone Waters, International House, 7 High Street, Ealing Broadway, London W5 5DW. www.volvic.co.uk. Tel. 44 20 8799 5800.

SUPPLIERS OF GOOD PROBIOTICS

OptiBac Probiotics, 15 Towergate Business Park, Colebrook Way, Andover, Hampshire SP10 3BB. www.optibacprobiotics.co.uk. Tel. 44 1264 339770. E-mail sales@mediapharma.co.uk

Biocare Ltd, Lakeside, 180 Lifford Lane, Kings Norton, Birmingham B30 3NU. www.biocare.co.uk. Tel. 44 121 433 3727. E-mail customerservice@biocare.co.uk

SUPPLIERS OF GOOD FAT AND PROTEIN

Laverstoke Farm – uk. North Overton, Overton, Hampshire RG25 3DR. www.laverstokepark.co.uk. Tel. 0800 394 5505

Ballymore Farm – Ireland. E-mail www.ballymorefarm.ie

SUPPLIERS OF GOOD HERBS

Bioforce Ltd, 2 Brewster Place, Irvine, KA11 5DD. www.avogel.co.uk. Tel: 1890 930070

SUPPLIERS OF FLAXSEED OIL

Virginia Harvest, Virginia Health Food Ltd, Oysterhaven, Kinsale, Co. Cork. www.virginiafoods.net. Tel. 353 21 4790033. E-mail info@virginiafoods.net

Udo's Choice, www.udoschoice.ie. Tel. 353 404 82444. E-mail customerservices@udoschoice.ie

SUPPLIERS OF VITAMINS AND MINERALS

Solgar, Beggars Lane, Aldbury, Tring, Herts HP23 5PT. www.solgar.com/uk. Tel. 44 1442 89055. E-mail solgarinfo@solgar.com

Biocare Ltd, Lakeside, 180 Lifford Lane, Kings Norton, Birmingham B30 3NU. www.biocare.co.uk. Tel. 44 121 433 3727. E-mail customerservice@biocare.co.uk

SUPPLIERS OF GREEN TEA

High Teas, www.highteas.co.uk

Index

Bold type indicates artwork/table/chart